Fundamentals of Aqueous Metallurgy

Fundamentals of Aqueous Metallurgy

Editor

Rakesh Shrivastav

Fundamentals of Aqueous Metallurgy

Edited by **Rakesh Shrivastav**

Printed in 2017

ISBN: 978-1-68117-485-3

Library of Congress Control Number: 2015936597

© 2016 by
SCITUS Academics LLC,
616, Corporate Way, Suite 2, 4766,
Valley Cottage, NY 10989

www.scitusacademics.com

Contents

vi

Preface

Extractive metallurgy is a branch of metallurgical engineering wherein process and methods of extraction of metals from their natural mineral deposits are studied. The field is a material science, covering all aspects of the types of ore, washing, concentration, separation, chemical processes and extraction of pure metal and their alloying to suit various applications, sometimes for direct use as a finished product, but more often in a form that requires further working to achieve the given properties to suit the applications. The field of ferrous and non-ferrous extractive metallurgy have specialties that are generically grouped into the categories of mineral processing, hydrometallurgy, pyrometallurgy, and electrometallurgy based on the process adopted to extract the metal. Several processes are used for extraction of same metal depending on occurrence and chemical requirements.

Editor

Chapter 1

Recovery of Ga(III) by Raw and Alkali Treated Citrus limetta Peels

Sachin C. Gondhalekar, and Sanjeev R. Shukla

Department of Fibres and Textile Processing Technology, Institute of Chemical Technology, Nathalal Parekh Marg, Matunga, Mumbai 400019, India

ABSTRACT

Alkali treated Citrus limetta peels were used for recovery of Ga(III) from its aqueous solution. The raw and alkali treated peels were characterized for functional groups. The efficiency of adsorption increased from 47.62 mg/g for raw peels to 83.33 mg/g for alkali treated peels. Between pH 1 and 3, the adsorption increased and thereafter decreased drastically. The adsorption followed pseudosecond order kinetics and Langmuir isotherm gave the best fit for the experimental data. Desorption studies showed 95.28% desorption after 3 cycles for raw peels while it was 89.51% for alkali treated peels. Simulated Bayer liquor showed 39.57% adsorption for gallium ions on raw peels which was enhanced to 41.13% for alkali treated peels.

INTRODUCTION

Gallium is the 30th in terms of abundance at an average concentration of 19 mg/g. It finds significant applications in the semiconductor industry through use of gallium arsenide and gallium nitride. Gallium is classified as a "strategic" metal, since it is used in high-tech gadgets like microwave transceivers, DVDs, laser diodes in CDs, and so forth and defence-related activities [1–3]. Gallium nitrate and gallium citrate are used in medical imaging as radio contrast agents [4]. Gallium arsenide is used to make rectifiers and amplifiers. Due to good adhesion to glass and high reflectivity, gallium is used in high quality mirrors [5]. There are no gallium-containing minerals of any economic significance. Due to its uniform distribution in soil, its extraction is uneconomical. It is usually associated with aluminium in bauxite, nephelines, and other ores and recovered as a by-product while producing alumina. The United States Geological Survey has estimated total world's primary gallium production to be about 273 tonnes in 2012 [6], the estimated consumption being 280 tonnes. As per the report of Roskill Information Services, neomaterial estimated that 50% of gallium consumed worldwide in 2010 came from recycled sources [7]. As per working Group Report of 12th Five Year Plan, India produced around 55 kg of gallium in recent years. Two plants, namely, Hindalco Industries Ltd. at Renukoot and National Aluminium Co. Ltd., at Damanjodi Alumina Refinery, Odisha, recover gallium [8]. It is derived from wastes of industrial processes, such as flue dusts from the zinc industry, waste generated during smelting of phosphate to produce elemental phosphorus, or sludge from the aluminium industry. Since aluminium lies in the same group of periodic table, gallium has affinity towards it. Bauxite (the primary aluminium ore) typically contains 0.003 to 0.01% gallium. Concentrations in zinc ores (e.g., sphalerite) are comparable. Since primary sources are scarce, the general strategy is to recover gallium from intermediate industrial products.

Bayer liquor is one of the important resource for recovery of gallium. About 90% of the world's primary gallium is produced from Bayer liquor [9]. In the Bayer process, about 70% of the gallium content of bauxite is leached along with aluminium and about 30% is retained in the red mud. Gallium accumulates in the Bayer liquor in successive cycles, attaining concentrations of 100–200 mg/L [10]. Gallium is also obtained from the iron mud or residues that result from the purification of zinc sulphate solutions, in zinc production. There are four kinds of recovery methods for gallium, namely, fractional precipitation, electrochemical deposition, solvent extraction, and ion exchange. Fractional precipitation is a complex process; solvent extraction is efficient but very slow; electrochemical method uses mercury cathode which is banned due to mercury toxicity, whereas ion exchange gives effective recovery of gallium. Industrial resins such as Duolite ES-346 and PHG586 may be used; however, the process is very expensive. Hence, further research for cheap, efficient, and environmentally friendly recovery of gallium is essential [11].

Biosorption studies have mainly focused on the removal and recovery of heavy metal ions from industrial effluents, detoxification being their primary goal. Functional groups present in the biosorbents that are responsible for metal binding are carboxyl, phosphoryl, sulfhydryl, amino, sulfate, imidazole, thioether, phenol, amide, and hydroxyl in various biomolecules of peptide, protein, and polysaccharide moieties of the cell walls. These functional groups bind the metal ions mainly by adsorption, ion exchange, and chelating effects [12].

Very few reports have been published on the removal/recovery of gallium using natural adsorbent [13–15].Citrus limetta peels are a natural pectinacious agrowaste material, typically generated in large quantities by the fruit juice industry, rich in pectin with abundant presence of carboxyl groups. It is reported that these peels have a metal binding mechanism similar to brown algal biosorbents since pectin is chemically similar to the brown algal cell wall polysaccharide alginate [6, 7]. These peels have been found to be very good adsorbents for Pb(II) and Cd(II) [16–18]. Our previous studies have shown good binding capacity by citrus peels for Pb(II) ions [18].

In the present work, the adsorption of Ga(III) from the aqueous solution of gallium nitrate and from the simulated Bayer liquor on Citrus limetta peels, in its raw as well as alkali treated form, has been reported.

MATERIALS AND METHODS

Materials

Citrus limetta fruit peels procured from local juice shop were used as biosorbents. Demineralized water was used throughout the experiment for dilution and washing purposes. Stock solution of 1000 mg/L was prepared from gallium nitrate salt (Sigma-Aldrich, India). Gallium standard used for calibration of Atomic Absorption Spectrometer (Model: GBC 932plus, Austria) was supplied by Merck Inc., Germany.

pH adjustment of metal ion solution was achieved by appropriate addition of 0.1 M NaOH and 0.1 M HCl.

METHODS

Preparation of Biosorbent

Citrus limetta peels procured from local juice shop were soaked in water for 4 h, washed, and cleaned using 1% nonionic detergent solution and again with demineralised water. The washed peels were dried at 60°C till constant weight, reduced to small particle size using a grinder, sieved to mesh sizes between 425 and 800 µ, and stored in zip lock bags. The biomass thus obtained was termed as raw citrus peels (RCP).

Modification of Functional Groups on RCP

The raw peels were treated with different chemical agents as reported to enhance their adsorption capacity. This data is further presented in Table 1 [19].

Table 1: Adsorption capacities of *Citrus limetta* peels (Ci = 100 mg/L, t = 4h, and T = 30°C) on chemical modification

Modification	Metal uptake (mg/L)
Nil (RCP)	24.40
1 M malonic acid treated peels	24.31
1 M oxalic acid treated peels	21.67
20% H_2O_2 treated peels	30.78
0.1 M sodium hydroxide treated peels (ACP)	34.59
1 M citric acid treated peels	23.90
1 M succinic acid treated peels	27.05
Acetylated peels	27.08
Chlorosulphonic acid treated peels	23.00
Methane sulphonic acid treated peels	23.33
0.1 M potassium dichromate treated peels	27.12

RCP was treated with 0.05 M aqueous NaOH for 3 h at room temperature. It was then washed thoroughly with demineralized water till neutral pH and then dried in an oven at 60°C for 24 h and stored in zip lock bags to avoid contamination with moisture. It was termed as alkali treated citrus peels (ACP). The weight loss of biomass due to alkali treatment was estimated experimentally.

Estimation of Gallium Ions

Ga(III) concentration in aqueous solution was determined by Atomic Absorption Spectrometer (GBC 932 plus, Australia) at 287.1 nm with an air-acetylene flame. Each time, AAS was calibrated by using standard 1000 mg/L Ga(III) solution.

ATR-IR Spectroscopy

The functional groups present on the surface of RCP and ACP were determined by recording the infrared spectra of these peels using Shimadzu 8400S FT-IR spectrometer (Figure 2).

Scanning Electron Microscopy (SEM)

Microscopic images of RCP and ACP were taken by scanning electron microscope (Jeol JSM 6380 LA spectrometer, Tokyo, Japan) to reveal the morphological aspects of the surface of the biomass samples.

Estimation of Acidic Groups

Number of acidic sites present on RCP and ACP were estimated by methylene blue absorption method [20,21]. When a biomass is treated with a cationic dye like methylene blue, the coloured cations of the dye are quantitatively absorbed by acidic groups present on the material forming strong ionic linkage. A weighed sample of the biomass was added to an Erlenmeyer flask containing 25 mL of aqueous methylene blue chloride solution (300 mg/L) and 25 mL borate buffer of pH 8.5. It was kept for 3 h at 25°C and then 5 mL of filtered sample was transferred to a volumetric flask containing 10 mL 0.1 M HCl followed by dilution to 100 mL. The decrease in the intensity of colour was measured on UV-visible spectrometer (Techcomp; UV-VIS 8500) at the $\lambda_{max} = 650$ nm. Using the calibration plot, the amount of unabsorbed methylene blue was calculated. The value gives quantitative information about the acidic sites present on the surface of peels.

Batch Wise Adsorption Experiments

Unless otherwise stated, for all the batch wise adsorption experiments, 100 mg of dry adsorbent was placed in 40 mL Ga(III) solution having varying initial concentrations ranging from 40 mg/L to 200 mg/L at pH 3.0 in a 250 mL stoppered Erlenmeyer flask and agitated in an orbital shaker (Rossari Biotech Ltd., Mumbai) at 150 rpm for 4 h at room temperature. The biomass was then allowed to settle down and the supernatant solution was pipetted out. The solutions were estimated for the remaining Ga(III) concentration using flame type AAS.

The biosorption capacity of gallium per unit of dry biomass (mg of Ga/g of dry biomass) was calculated by using

$$q_{eq} = \frac{\left(C_0 - C_{eq}\right) V}{W},$$

(1)

where q_{eq} is the equilibrium adsorption capacity (mg/g), C_0 and C_{eq} are the initial and equilibrium concentrations of Ga(III) (mg/L), respectively, V is the volume of adsorbate (L), and W is the mass of adsorbent (g).

The biosorption efficiency, E%, of the metal ion was calculated from

$$E\% = \frac{\left(C_0 - C_{eq}\right)}{C_0} \times 100.$$

(2)

The optimum pH for the biosorption was evaluated by adjusting the pH between 1 and 3 with an interval of 0.5 using 1 M HCl.

The kinetics of adsorption of Ga(III) was studied by shaking 2.5 g of adsorbent samples with 200 mL of approximately 100 mg/L solutions of Ga(III) at pH 3.0 and at room temperature (30°C) up to 300 min in a 250 mL Erlenmeyer flask at 150 rpm. At each predetermined time, 1 mL of the solution was removed and tested for its metal content using flame type AAS. The amount of metal ions adsorbed was evaluated from the difference between the initial and final concentrations of Ga(III) ions in the solution.

Isotherm Modeling

Out of many isotherm models, the Langmuir and Freundlich models are the most commonly studied due to their ease of interpretation [22]. Langmuir isotherm was initially derived for adsorption of gases on solid surfaces (Figure 6), and it considered sorption as a chemical phenomenon; the sorbent surface contains only one type of binding site and the sorption is limited to monolayer [23]. It is given by

$$q_{eq} = \frac{K_l q_{max} C_{eq}}{1 + K_l C_{eq}},$$

(3)

where q_{eq} is the metal uptake, q_{max} is the maximum biosorption capacity, K_l is the constant related to adsorption energy, and C_{eq} is the equilibrium concentration of metal ions. The Langmuir parameters can be determined from the slope and the intercept of the plot C_{eq}/q_{eq} versus C_{eq}, based on the linearized form of the above equation, can be written as

$$\frac{C_{eq}}{q_{eq}} = \frac{1}{K_l q_{max}} + \frac{C_{eq}}{q_{max}}.$$

(4)

Freundlich proposed an empirical isotherm relation as expressed in

$$q_{eq} = K_f C_{eq}^{1/n},$$

(5)

in which K_f and n are Freundlich constants (Figure 7). As the Freundlich isotherm is exponential, it can be reasonably applied only in the low to intermediate concentration range [24]. The equation can be linearized as

$$\log q_{eq} = \log K_f + \frac{1}{n} \log C_{eq}.$$

(6)

Besides these two common adsorption isotherms, Sips isotherm is also widely used to study the biosorption mechanism of pectin containing compounds [25] (Figure 8). This isotherm is a combination of Langmuir and Freundlich isotherm equations, deduced for predicting the heterogeneous adsorption systems and to overcome the limitation of the rising adsorbate concentration associated with Freundlich isotherm model. At low adsorbate concentrations, it reduces to Freundlich isotherm; while, at high concentrations, it predicts a monolayer adsorption capacity characteristic of the Langmuir isotherm (Tables 4 and 5). As a general rule, the equation parameters are governed mainly by the operating conditions such as the alteration of pH, temperature, and concentration. Sips isotherm can be expressed as

$$q_{eq} = q_{max} \frac{K_s C_{eq}^{n_s}}{1 + K_s C_{eq}^{n_s}},$$

(7)

in which n_s is the Sips constant [26] (Table 6).

Kinetic Modeling

Biosorption data under nonequilibrium conditions is usually described by pseudofirst order and pseudosecond order equations [27].

Pseudofirst Order Equation. The pseudofirst order kinetic equation or Lagergren equation is given as

$$\frac{dq_t}{dt} = k_1 \left(q_{eq} - q_t\right),$$

(8)

in which q_t is the amount of adsorbate adsorbed at time t, q_{eq} is the value at equilibrium, and k_1 is the constant [28].

Pseudosecond Order Equation. The pseudosecond order kinetic equation has been frequently employed to analyze the biosorption data using different adsorbates and biosorbents as reviewed by Ho and McKay [29]. Consider

$$\frac{dq_t}{dt} = k_2 \left(q_{eq} - q_t\right)^2,$$

(9)

in which k_2 is a constant.

Adsorption-Desorption Cycles

Repetitive adsorption-desorption studies were carried out to evaluate the economic feasibility of the process. Desorption of Ga(III) from previously adsorbed peels was carried out by shaking them with 40 mL of desorbing media at 30°C for 240 min. Different desorbing media with varied concentrations, studied for the desorption, were HCl, HNO_3, and H_2SO_4. The metal ion content in the desorbing media was then estimated using AAS.

Regeneration of Biosorbent

RCP and ACP were thoroughly washed with demineralized water after desorption, treated with 40 mL of 0.025 M NaOH solution for 240 min, washed again, dried at 60°C in hot air oven for 24 h, and reused for another adsorption-desorption cycle.

Preparation of Simulated Bayer Liquor

The process feasibility was checked by adsorbing the Ga(III) ions from the simulated spent Bayer liquor on RCP and ACP. It was prepared in laboratory by adding the components which are generally present in spent Bayer liquor in their stoichiometric amount [30]. Thus, 25 g of Na_2CO_3 were dissolved in hot water and cooled to room temperature. To this, 125 g of NaOH pellets were added slowly with stirring followed by addition of 120 g of $Al(OH)_3$. The solution was heated near to boiling point, cooled, filtered, and diluted to 1 L. Then 14.7 mL of gallium stock solution (13.62 g/L) was transferred into the solution and the solution was made up to 1 L by adding water. Forty mL of this simulated spent Bayer liquor was taken in an Erlenmeyer flask, with pH adjusted to 3.0 by 0.1 M HCl, and subjected to adsorption on 0.1 g of biomass. The concentration of Ga(III) ions before and after the adsorption was estimated on AAS. Weight loss of the biomaterials during adsorption process was also estimated gravimetrically.

RESULTS AND DISCUSSION

Effect of Pretreatment on Ga(III) Biosorption

Citrus peels are mainly made up of pectin (around 30% by mass), which is rich in galacturonic acid [31]. In nature, around 80% of carboxyl groups of galacturonic acid are present in esterified form as methyl carboxylate. The nonesterified galacturonic acid units can be either free acids (carboxyl groups) or salts with sodium, potassium, or calcium. The salts of partially esterified pectins are called pectinates. Salts below 5% degree of esterification are called pectates and the insoluble acid form is the pectic acid [32]. Alkali treatment is very effective in hydrolysis of these esterified galacturonic acid units and converts them into free acid sites. It also ruptures the cell wall and exposes more functional groups, thereby promoting the heavy metal adsorption by preferential ion-exchange mechanism [33]. Earlier oxidative pretreatment of cellulosic biomass such as jute and coir with hydrogen peroxide has shown to enhance the adsorption capacity for Pb(II) cations [34, 35]. Oxidized coir also showed enhanced adsorption of gallium (19.42 mg/g) compared to unmodified coir (13.75 mg/g) [13].

Raw Citrus limetta peels (RCP), a pectin containing waste biomaterial, have shown good potential to adsorb Pb(II) ions [18]. Initial experiments on RCP also showed the potential to adsorb Ga(III) ions from its aqueous solution (C_{eq} = 21.70 mg/g for C_i = 70 mg/L). RCP was given various chemical treatments that are reported to enhance the adsorption capacity of biomaterials [19] (Table 1).

Among those, it was found that the alkali treatment was the most efficient in enhancing the adsorption capacity of peels and hence the concentration of NaOH was optimized at RT (30°C) and 2 h treatment time (Figure 1) (Table 2). Different acidic groups such as carboxylic and sulphonic acid also get converted into their sodium forms which have been shown to promote heavy-metal ion adsorption [36].

Table 2: Effect of concentration of NaOH on weight loss of biosorbent

NaOH concentration (N)	Weight loss (%)
0.000	00.0
0.001	19.0
0.010	19.2
0.050	19.6
0.100	20.8

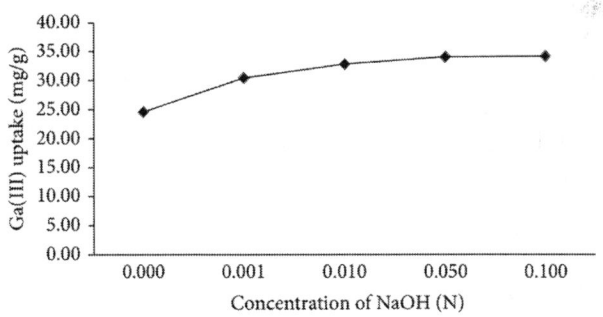

Figure 1: Effect of different concentration of NaOH on biosorption capacity of *Citrus limetta* peels (C_i = 100mg/L, t = 4h, T = 30°C).

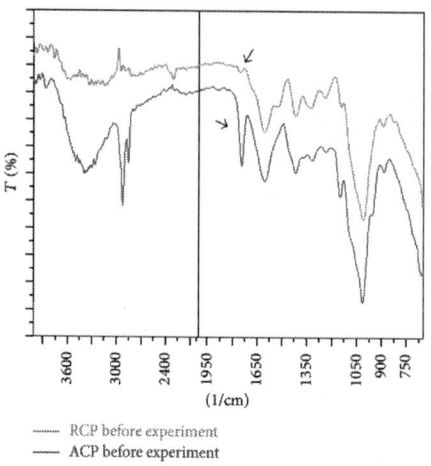

------- RCP before experiment
------- ACP before experiment

Figure 2: IR plot of RCP and ACP.

FT-IR Spectra of Biomass

FTIR spectra of peels showed peak at ~3350 cm^{-1} which is characteristic of hydroxyl group, mainly due to water. The peak observed at ~1625 cm^{-1} is attributed to asymmetric stretching of the carboxylic (C=O) double bond. The peaks observed at 2924 cm^{-1} and 2856 cm^{-1} in ACP are due to asymmetric and symmetric stretching modes of methylene groups. No strong shift in the wave number ~1750 cm^{-1} was observed in RCP, although the intensity of peak increased in ACP which is characteristic of carbonyl group of carboxylic acid (–COOH) and ester (–COOR). This indicates increase in the number of carboxylic groups in ACP which has also been quantified by methylene blue absorption method to 6.32×10^{-2} mmol/g from 5.40×10^{-2} mmol/g in RCP. Thus, the alkali treatment caused modification of the ester functional groups to carboxylic acid groups present in raw peels.

Scanning Electron Microscopy

SEM images clearly indicate the morphological changes in the peels on alkali treatment. It can be easily depicted from the SEM that the surface of peels is highly heterogeneous. As seen in Figure 3, ACP surface shows more cavities and more number of opened up pores as compared to RCP indicating that it has increased the surface area available for adsorption. Alkali treatment proved to be effective in rupturing the cell walls due to which more functional groups are exposed on the surface and became available for adsorption. Apart from that, no further significant morphological changes were apparent in the SEM images.

Figure 3: Scanning electronic micrographs of (a) RCP and (b) ACP with magnification 500x.

Estimation of Acidic Groups on the Surface of Peels

The carboxyl groups present in the biosorbents have been proved to be directly responsible for the sorption of heavy metals [17, 18]. These are the most abundant acidic functional groups and the adsorption capacity of peels is directly related to the presence of these sites in pectin in the form of galacturonic acid. Estimation of acidic groups on the surface of peels available for sorption reveals that ACP contains contains more carboxylic acid groups (6.32×10^{-2} mmol/g) as compared to RCP (5.40×10^{-2} mmol/g). This has been attributed to the alkaline hydrolysis of some of the ester groups present on the surface of the peels under the treatment conditions. Structure of pectin before and after hydrolysis is as shown in Figure 4.

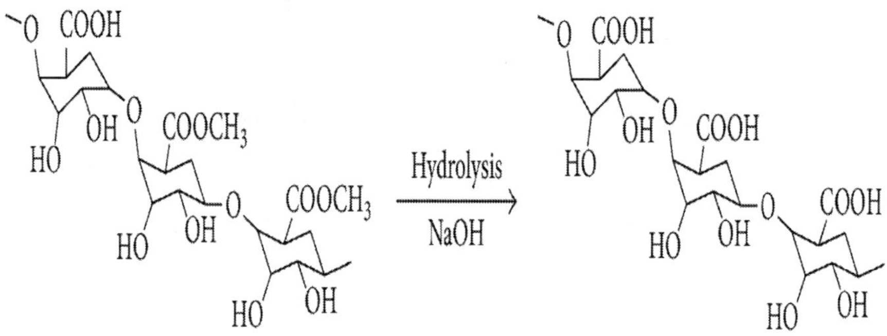

Figure 4: Structure of pectin before and after alkaline treatment.

Effect of Initial Solution pH

pH is one of the most crucial factors which drives the efficiency of adsorption, as it governs the speciation of the metal ions in aqueous solution and also determines the degree of protonation on the biomass. At lower pH, protons compete with the metal ions thereby decreasing the adsorption, while, at higher pH, metal ions form corresponding hydroxides get precipitated out.

The biosorption of Ga(III) was strongly affected by initial solution pH. The sorption capacity of RCP has increased from 16.29 mg/g at pH 1.0 to 35.59 mg/g Ga(III) at pH 3.0 (Figure 5) indicating an increase in adsorption capacity. Beyond pH 3.0, precipitation occurs, as gallium forms a hydroxide gel and loses its solution characteristics [36]. In case of ACP, the increase in adsorption capacity registered was from 18.23 mg/g to 44.44 mg/g at similar pH values.

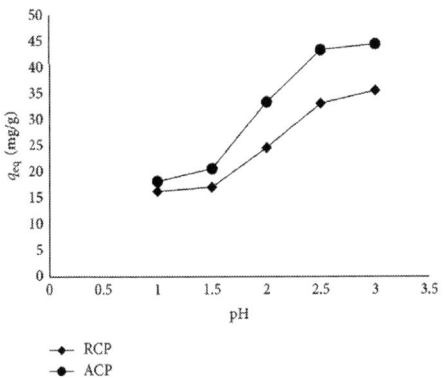

Figure 5: Effect of pH on biosorption of Ga(III).

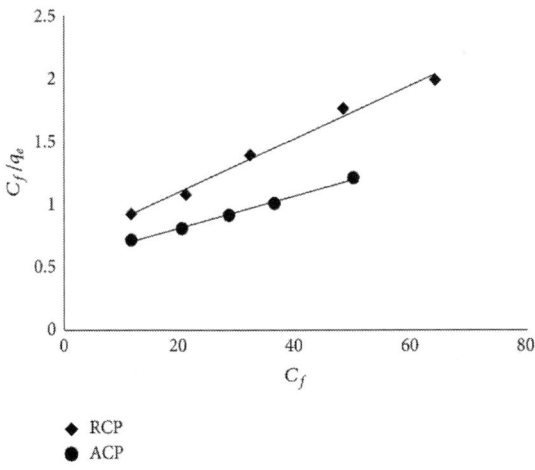

Figure 6: Linearized Langmuir isotherm.

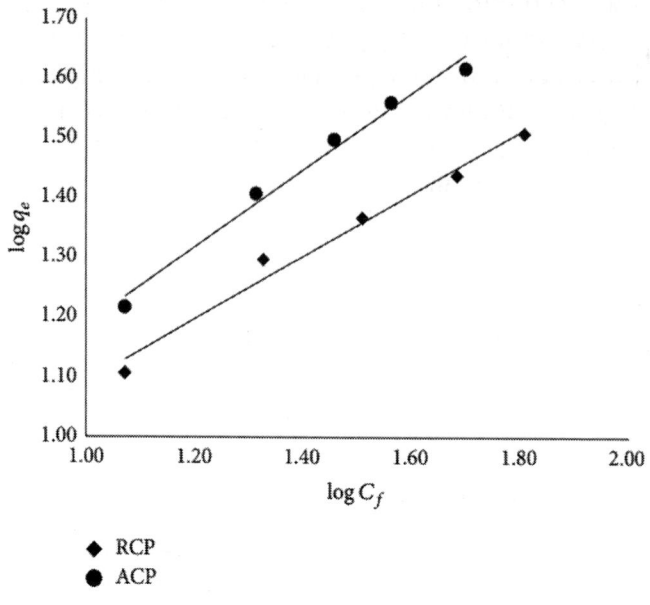

Figure 7: Linearized Freundlich isotherm.

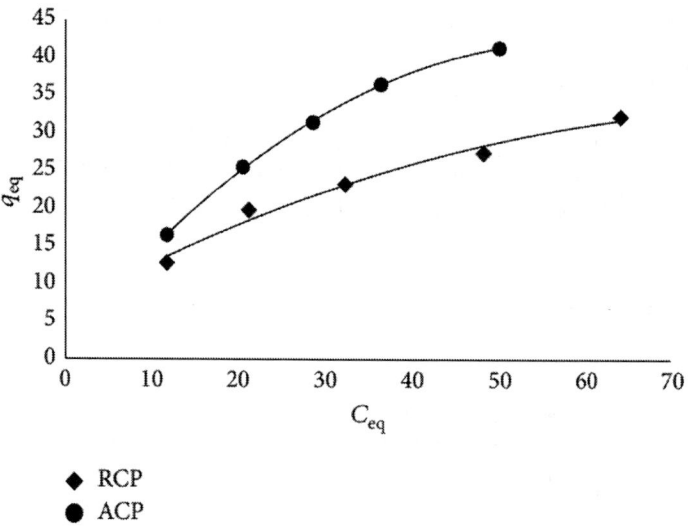

Figure 8: Nonlinear Sips isotherm.

In aqueous solution, gallium is always present in its hydrated form with six molecules of water held strongly to make an octahedral complex. The strength of metal-oxygen bond weakens O–H bond; hence, hydrolysis occurs and protons are released, thus giving acidic solution. The more metal ion concentration in the solution is, the more the hydrolysis is and, hence, the more acidic the solution becomes [37, 38]. Ga(III) ions are Lewis acids, and, in aqueous solution, they form aqua ions of the formula $Ga(H_2O)_6$

$^{3+}$ The aqua ions undergo hydrolysis; the hydrated gallium ions have six molecules of water which are held firmly giving an octahedral complex. The first hydrolysis step is given generically as

Thus, the aqua cations behave as acids in terms of Brønsted-Lowry acid-base theory. This effect is easily explained by considering the inductive effect of the positively charged metal ion, which weakens the O–H bond of an attached water molecule, causing the liberation of a proton relatively easy to make the solution acidic.

Like other metal ions, the adsorption of Ga(III) is also highly influenced by pH [39–41]. At lower pH, the affinity towards the proton of the binding site of peels is much greater than that of the metal ion ($H^+ \gg M^{3+}$), compared with that at higher pH where $M^{3+} \gg H^+$ [36]. Therefore, the adsorption is less at a lower pH of 1.0 which remains almost unaffected till pH 1.5 as Ga(III) has strong competition with protons and then it increases with the pH as the concentration of protons decreases till pH 2.5 is attained.

From the data given in Table 3, it is clear that the adsorption of Ga(III) on Citrus limetta peels is much higher than any other biosorbent used in earlier studies [13]. This may be attributed to enhanced carboxylic acid sites in ACP pectin.

Table 3: Effect of biosorbent on adsorption of Ga(III) from aqueous solution

Biosorbent used	Initial Ga concentration (mg/L)	q_{max} (mg/g)	Reference
Oxidized coir	203	19.42	[13]
Alkali treated peels	200	76.26	This study

Table 4: Values obtained from Langmuir and Freundlich isotherms (linear parameters)

Adsorbent	Langmuir isotherm				Freundlich isotherm			
	q_{max} (mg/g)	b	r^2	RMSE	K_f	n	r^2	RMSE
RCP	47.62	0.031	0.99	3.02	3.84	1.61	0.99	7.08
ACP	83.33	0.022	0.99	3.69	3.69	1.90	0.98	7.72

Table 5: Values obtained from Langmuir and Freundlich isotherms (nonlinear parameters)

Adsorbent	Langmuir isotherm		Freundlich isotherms	
	q_{max} (mg/g)	b	K_f	n
RCP	46.54	0.033	4.04	0.50
ACP	76.26	0.024	4.11	0.60

Table 6: Values obtained from Sips isotherms

Adsorbent	Sips isotherm			
	K_s	β_s	a_c	r^2
RCP	1.124	1.212	0.018	0.99
ACP	2.412	0.787	0.038	0.99

Adsorption Isotherm

The adsorption data were fitted to Langmuir and Freundlich adsorption isotherms. The metal loading capacities (q_{max}) of RCP and ACP as calculated from the slope of the plot C_{eq}/q_{eq} versus C_{eq} were found to be 47.62 and 83.33mg/g, respectively. Though the data seems to be fitting equally well with the Langmuir ($r^2 = 0.99$) and Freundlich ($r^2 = 0.99$) isotherms, RMSE error value for Langmuir plot is much lower than Freundlich plot. This suggests that the Langmuir model gives better fit for both the adsorbents. "K_f" and "n" were calculated from the intercept and slope of the plot $\log q_{eq}$ versus $\log C_{eq}$. Freundlich isotherm values of "n" were 1.61 and 1.90 for RCP and ACP, respectively, suggesting favourable adsorption by ACP.

Kinetic Modelling

Pseudosecond order (R^2 = 0.99) plot shows better fit than pseudofirst order (R^2 = 0.94) for RCP and ACP. Figure 9 represent the linearized plots for second order model for RCP and ACP. Higher value of initial rate, h, in ACP (2.50mg/g·min) suggests favorable adsorption. The values for pseudosecond order model has been listed in Table 7. The kinetic study reveals closeness of the predicted and experimental q_{max} values, which are 25.00 and 24.11 for RCP and 32.26 and 29.77 for ACP, respectively. Pseudosecond order plot gives best fit for RCP as well as ACP; hence, it was considered that adsorption of Ga(III) on RCP and ACP takes place via pseudosecond order kinetics.

Table 7: Experimental and pseudosecond order values for RCP and ACP

Adsorbent	Experimental q_{eq} (mg/g)	k^2($\times10^{-3}$/min) (/min)	q_{eq} (mg/g)	r^2	Initial rate, h(mg/g·min)
RCP	24.11	2.7	25.00	0.99	1.69
ACP	32.26	2.4	29.77	0.99	2.50

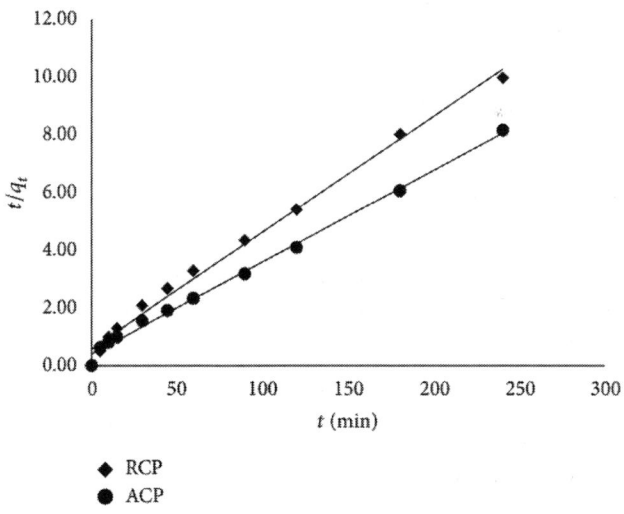

Figure 9: Pseudosecond order plot for C_0 = 100mg/L (RCP) and C_0= 100mg/L (ACP).

Desorption Studies

Desorption efficiency of the adsorbent is crucial to the recovery of gallium. The metal ion loaded adsorbent was most effectively desorbed by using 0.5 M HCl, when different acids (HCl, H_2SO_4, and HNO_3) with different concentrations (0.1 M–1.0 M) were tried. Ga(III), adsorbed on the surface of the adsorbent, exchanges with the H^+ ion of the acid due to its high affinity towards the functional groups, behaving as a cation exchanger. Desorption in the first cycle was observed to be more than 95%, which decreased marginally in the subsequent cycles. The weight loss study of the adsorbents during the adsorption-desorption cycles indicated that, after the first complete cycle, the weight loss for RCP was 19.19% while that for ACP was 17.53%. It increased to 37.05% for RCP and to 55.45% for ACP after the second cycle. Even after the third cycle, the same adsorbent can be used by seeing its good adsorption capacity.

Regeneration Studies

Although the Citrus limetta peels is a waste material obtained from the orange fruit, any chemical treatment to convert it into a better adsorbent adds to the cost. In order to make the process economically viable, the peels after desorption of Ga(III) were regenerated using 0.025N NaOH and reused for another adsorption-regeneration cycle. Low concentration of alkali is used as regeneration media for economic viability. Regeneration of the peels was studied for 3 cycles and showed comparative adsorption in the next cycles.

Simulated Bayer Liquor Sample Analysis

Synthetic Bayer liquor sample was analyzed for the efficiency of biosorbent for adsorption-desorption cycles (Table 9). It was observed that RCP showed 39.57% adsorption of gallium from the simulated spent Bayer liquor while ACP showed 41.13% adsorption. Both the adsorbents showed nearly complete desorption in 0.5 M HCl as desorbing media. The weight loss of biomass after first cycle was observed to be 29.98% for RCP and 18.03% for ACP. The decrease in the adsorption capacity as compared to pure aqueous solution of gallium may be attributed to the presence of other metal ions such as Na(I) and Al(III) present in the solution. These results are presented in Table 8. It may be seen that RCP and ACP are competitive to synthetic ion exchange resin showing adsorption capacities of 7.32 and 7.61 mg/g, respectively, which are higher than that of reported value (6.0 mg/g) on resin with hydroxamic acid ligand [3].

Table 8: Repeated adsorption-desorption cycles for Ga(III) adsorption

Adsorbent	Cycle 1			Cycle 2				Cycle 3			
	% adsorption	% desorption	% weight loss	% adsorption	% desorption	% weight loss	Overall weight loss	% adsorption	% desorption	% weight loss	Overall weight loss
RCP	58.98	98.71	19.19	34.37	95.77	16.21	39.40	37.08	95.26	4.76	40.16
ACP	68.51	95.37	17.53	56.00	89.78	12.90	30.43	55.46	89.51	2.88	33.31

Table 9: Synthetic Bayer liquor adsorption-desorption cycle

Adsorbent	Adsorption (%)	Desorption (%)	Weight loss (%)
RCP	39.57	98.35	29.98
ACP	41.13	92.23	18.03

CONCLUSIONS

Waste biomass Citrus limetta peels were effectively used to adsorb gallium ions from aqueous solution. When treated with sodium hydroxide, the alkali treated peels (ACP) showed enhanced metal removal capacity. Optimum pH for adsorption was found to be 3.0. Kinetic data showed that the equilibrium reached within 180 minutes. Pseudosecond order model proved to be the best fit whereas isotherm studies reveal that the metal ion adsorption capacity of 47.62 mg/g for RCP was enhanced to 83.33 mg/g for ACP. The adsorbents showed good stability up to three adsorption-desorption cycles and hence proved practical to use at scale up level. Also RCP and ACP showed significant adsorption in synthetic Bayer liquor sample which highlights the use ofCitrus limetta peels at commercial level.

ACKNOWLEDGMENTS

The author Sachin C. Gondhalekar gratefully acknowledges the funding by the University Grants Commission's fellowship under Special Assistance Programme.

REFERENCES

1. R. R. Moskalyk, "Gallium: the backbone of the electronics industry," Minerals Engineering, vol. 16, no. 10, pp. 921–929, 2003. · ·

2. I. R. Grant, "Gallium arsenide from mine to microcircuit," Transactions of the Institution of Mining and Metallurgy, C: Mineral Processing and Extractive Metallurgy, vol. 97, pp. 48–52, 1987.

3. P. Selvi, M. Ramasami, M. H. P. Samuel, P. Adaikkalam, and G. N. Srinivasan, "Recovery of gallium from Bayer liquor using chelating resins in fixed-bed columns," Industrial and Engineering Chemistry Research, vol. 43, no. 9, pp. 2216–2221, 2004.

4. G. Iosilevsky, D. Front, L. Bettman, R. Hardoff, and Y. Ben-Arieh, "Uptake of gallium-67 citrate and [2-3H]deoxyglucose in the tumor model, following chemotherapy and radiotherapy," Journal

of Nuclear Medicine, vol. 26, no. 3, pp. 278–282, 1985.

5.   E. F. Borra, R. Content, L. Girard, S. Szapiel, L. M. Tremblay, and E. Boily, "Liquid mirrors: optical shop tests and contributions to the technology," Astrophysical Journal Letters, vol. 393, no. 2, pp. 829–847, 1992.

6.   U.S. Geological Survey, Mineral Commodity Summaries, 2013.

7.   A 2011 Roskill Report, "Gallium: global industry markets & outlook," Roskill Report, 2011.

8.   Govt of India, Ministry of Mines, Indian Beauro of Mines, Indian Mineral Year Book, vol. 2, 2011.

9.   X. Lu, L. Wang, X. Wang, and X. Niuesearch, "Research progress in gallium recovery technology,"Nonferrous Metals, vol. 60, pp. 105–108, 2008.

10. A. M. G. Figueiredo, W. Avristcher, E. A. Masini, S. C. Diniz, and A. Abrão, "Determination of lanthanides (La, Ce, Nd, Sm) and other elements in metallic gallium by instrumental neutron activation analysis," Journal of Alloys and Compounds, vol. 344, no. 1-2, pp. 36–39, 2002. · ·

11. Z. Zhao, Y. Yang, Y. Xiao, and Y. Fan, "Recovery of gallium from Bayer liquor: a review,"Hydrometallurgy, vol. 125-126, pp. 115–124, 2012. · ·

12. J. Wang and C. Chen, "Biosorbents for heavy metals removal and their future," Biotechnology Advances, vol. 27, no. 2, pp. 195–226, 2009. · ·

13. U. S. Suryavanshi and S. R. Shukla, "Adsorption of Ga(III) on oxidized coir," Industrial and Engineering Chemistry Research, vol. 48, no. 2, pp. 870–876, 2009.

14. H. Eroglu, S. Yapici, C. Nuhoglu, and E. Varoglu, "Biosorption of Ga-67 radionuclides from aqueous solutions onto waste pomace of an olive oil factory," Journal of Hazardous Materials, vol. 172, no. 2-3, pp. 729–738, 2009. · ·

15. H. Eroglu, S. Yapici, and E. Varoglu, "An investigation of the biosorption of radioactive gallium-67 in an aqueous solution using rose residue," Journal of Chemical and Engineering Data, vol. 55, no. 8, pp. 2848–2856, 2010. · ·

16. S. Schiewer and A. Balaria, "Biosorption of Pb^{2+} by original and protonated citrus peels: equilibrium, kinetics, and mechanism,"

Chemical Engineering Journal, vol. 146, no. 2, pp. 211–219, 2009. · ·

17. E. Njikam and S. Schiewer, "Optimization and kinetic modeling of cadmium desorption from citrus peels: a process for biosorbent regeneration," Journal of Hazardous Materials, vol. 213-214, pp. 242–248, 2012. · ·

18. U. Suryavanshi and S. R. Shukla, "Adsorption of Pb^{2+} by Alkali-treated citrus limetta peels," Industrial and Engineering Chemistry Research, vol. 49, no. 22, pp. 11682–11688, 2010. · ·

19. S. Srivastava and P. Goyal, Novel Biomaterials: Decontamination of Toxic Metals from Wastewater, Springer, 2010.

20. G. Sèbe, P. Pardon, F. Pichavant, S. Grelier, and B. De Jéso, "An investigation into the use of eelgrass (Zostera noltii) for removal of cupric ions from dilute aqueous solutions," Separation and Purification Technology, vol. 38, no. 2, pp. 121–127, 2004. · ·

21. D. Klemm, B. Philipp, T. Heinze, U. Heinze, and W. Wagenknecht, "Fundamentals and analytical methods," in Comprehensive Cellulose Chemistry, vol. 1, p. 236, John Wiley & Sons, Weinheim, Germany, 1998.

22. K. Y. Foo and B. H. Hameed, "Insights into the modeling of adsorption isotherm systems," Chemical Engineering Journal, vol. 156, no. 1, pp. 2–10, 2010. · ·

23. I. Langmuir, "The adsorption of gases on plane surfaces of glass, mica and platinum," The Journal of the American Chemical Society, vol. 40, no. 9, pp. 1361–1403, 1918.

24. A. Z. Freundlich, "Über die adsorption in lösungen," Zeitschrift fur Physikalische Chemie, vol. 57, pp. 385–470, 1906.

25. K. V. Kumar and K. Porkodi, "Relation between some two- and three-parameter isotherm models for the sorption of methylene blue onto lemon peel," Journal of Hazardous Materials, vol. 138, no. 3, pp. 633–635, 2006. · ·

26. R. Sips, "Combined form of langmuir and freundlich equations," Journal of Chemical Physics, vol. 16, pp. 490–495, 1948.

27. Y. Liu and Y.-J. Liu, "Biosorption isotherms, kinetics and thermodynamics," Separation and Purification Technology, vol. 61, no. 3, pp. 229–242, 2008. · ·

28. S. Lagergren, "Zur theorie der sogenannten adsorption gelöster stoffe. Kungliga Svenska Vetenskapsakademiens," Handlingar, vol. 24, no. 4, pp. 1–39, 1898.

29. Y. S. Ho and G. McKay, "Pseudo-second order model for sorption processes," Process Biochemistry, vol. 34, no. 5, pp. 451–465, 1999. · ·

30. F.-F. Zha, A. G. Fane, and C. J. D. Fell, "Liquid membrane processes for gallium recovery from alkaline solutions," Industrial and Engineering Chemistry Research, vol. 34, no. 5, pp. 1799–1809, 1995.

31. C. D. May, "Industrial pectins: sources, production and applications," Carbohydrate Polymers, vol. 12, no. 1, pp. 79–99, 1990.

32. P. Sriamornsak, "Chemistry of pectin and its pharmaceutical uses: a review," Journal of Pharmacy and Pharmaceutical Sciences, vol. 44, pp. 207–228, 1998.

33. S. R. Shukla, R. S. Pai, and A. D. Shendarkar, "Adsorption of Ni(II), Zn(II) and Fe(II) on modified coir fibres," Separation and Purification Technology, vol. 47, no. 3, pp. 141–147, 2006. · ·

34. S. R. Shukla and R. S. Pai, "Comparison of Pb(II) uptake by coir and dye loaded coir fibres in a fixed bed column," Journal of Hazardous Materials, vol. 125, no. 1–3, pp. 147–153, 2005. · ·

35. S. R. Shukla and R. S. Pai, "Adsorption of Cu(II), Ni(II) and Zn(II) on modified jute fibres," Bioresource Technology, vol. 96, no. 13, pp. 1430–1438, 2005. · ·

36. P. Persson, K. Zivkovic, and S. Sjöberg, "Quantitative adsorption and local structures of gallium(III) at the water--FeOOH interface," Langmuir, vol. 22, no. 5, pp. 2096–2104, 2006. · ·

37. J. Burgess, Metal Ions in Solution, Ellis Horwood, New York, NY, USA, 1978.

38. D. T. Richens, The Chemistry of Aqua Ions: Synthesis, Structure, and Reactivity: A Tour Through the Periodic Table of the Elements, John Wiley & Sons, 1997.

39. N. Fiola, I. Villaescusaa, M. Martínezb, N. Mirallesb, J. Pochc, and J. Serarols, "Sorption of Pb(II), Ni(II), Cu(II), and Cd(II) from aqueous solution by olive stone waste," Separation and Purification Technology, vol. 50, no. 1, pp. 132–140, 2006.

40. R. H. S. F. Vieira and B. Volesky, "Biosorption: a solution to pollution?" International Microbiology, vol. 3, no. 1, pp. 17–24, 2000.

41. B. Volesky, Biosorption of Heavy Metals, vol. 3, CRC Press, Boca Raton, Fla, USA, 1990.

Chapter 2

Direct Formation of Gold Nanoparticles on Substrates Using a Novel ZnO Sacrificial Templated-growth Hydrothermal Approach and their Properties in Organic Memory Device

Lean Poh Goh[1], Khairunisak Abdul Razak[1, 2], Nur Syafinaz Ridhuan[1, 2], Kuan Yew Cheong[1], Poh Choon Ooi[3], and Kean Chin Aw[3]

[1]School of Materials and Mineral Resources Engineering, Universiti Sains Malaysia, Nibong Tebal, Penang 14300, Malaysia

[2]NanoBiotechnology Research and Innovation, INFORMM, Universiti Sains Malaysia, USM, Penang, 11800, Malaysia

[3]Mechanical Engineering, The University of Auckland, Auckland, 1142, New Zealand

ABSTRACT

This study describes a novel fabrication technique to grow gold nanoparticles (AuNPs) directly on seeded ZnO sacrificial template/ polymethylsilsesquioxanes (PMSSQ)/Si using low-temperature hydrothermal reaction at 80°C for 4 h. The effect of non-annealing and various annealing temperatures, 200°C, 300°C, and 400°C, of the ZnO-seeded template on AuNP size and distribution was systematically studied. Another PMMSQ layer was spin-coated on AuNPs to study the memory properties of organic insulator-embedded AuNPs. Well-distributed and controllable AuNP sizes were successfully grown directly on the substrate, as observed using a field emission scanning electron microscope followed by an elemental analysis study. A phase analysis study confirmed that the ZnO sacrificial template was eliminated during the hydrothermal reaction. The AuNP formation mechanism using this hydrothermal reaction approach was proposed. In this study, the AuNPs were charge-trapped sites and showed excellent memory effects when embedded in PMSSQ. Optimum memory properties of PMMSQ-embedded AuNPs were obtained for AuNPs synthesized on a seeded ZnO template annealed at 300°C, with 54 electrons trapped per AuNP and excellent current–voltage response between an erased and programmed device.

BACKGROUND

Organic materials and devices have drawn attention for applications in modern electronic devices due to their excellent process ability in large-area circuits; possibility for molecular design through chemical synthesis; high mechanical flexibility, comparable to flexible substrates; low processing cost; lower power consumption; good scalability; multiple state property; three-dimensional stacking capability; and large data storage capacity [1-5]. Therefore, organic memories have aroused wide interest for electronic devices in new information technology. Organic memory devices can be divided into several device structures as follows: organic capacitors, organic field-effect transistors, organic diodes, and metal/organic semiconductor/metal junctions. Organic memory devices can further be divided based on the charge storage mechanism as follows: ferroelectric, polymer charge trapping, floating-gate storages, and those with nanoparticle use [5].

Organic memory has recently drawn increasing attention. Using a three-layer stacking structure, a tri-layer (polymer/metallic nanoparticle/polymer) structure has demonstrated a bistable memory effect. This simple tri-layer can be used to construct a memory device in a structure consisting of metal/tri-layer/semiconductor layers, the MIS structure. The tri-layer is most important because it is where charge trapping occurs. The memory function of the trilayer structure can be achieved by storing charge in nanoparticles or nanoclusters, which are interposed between the insulating polymer layers. The semiconductor layer is used as an electronic charge source to be injected into the tri-layer to be trapped by the nanoparticles [6-9]. Among the metallic nanoparticles, gold nanoparticles (AuNPs) possess important properties for device fabrication as well as good memory characteristics, such as easy synthesis approach, high work function [10-12], good electron-accepting properties[13], and chemical stability [12]. A broad work function improves the retention time and speed of the write-erase process [14].

The formation and properties of AuNP thin films embedded in organic memory have been investigated using various methods including typical reduction method followed by spin coating [15,16], Langmuir-Blodgett film deposition [17], dip coating [18], and template-directed assembly method [19]. For AuNPs synthesized using the reduction method followed by spin coating to form thin films, AuNPs were normally non-uniformly distributed. Therefore, more AuNPs can be found at the edges of a substrate due to the centrifugal effect, resulting in non-uniform AuNP dispersion that leads to inconsistent memory device properties. AuNPs synthesized using the Langmuir-Blodgett film deposition require complicated processing steps and involve several chemical reactions. Kim et al.[18] used 3-aminopropyl-triethoxysilane to modify a surface in the self-assembly of AuNPs to produce uniform and stable AuNP adsorption on the dielectric surface. Template use allows the controllable formation of AuNPs with specific geometric characteristics. Meanwhile, the template-directed assembly method is promising for obtaining well-distributed AuNPs for a memory device. However, the polymer template has to be removed using O_2 plasma [19].

The current study presents a novel method in forming AuNPs directly on a substrate. The AuNPs were grown on a seeded ZnO template using the low-temperature sacrificial hydrothermal growth technique.

The seeded template refers to the nanosize grain-consisting ZnO layer after annealing at a certain temperature. In a typical ZnO nanorod formation, this ZnO seed template contributes to the formation of well-aligned nanorod arrays [20, 21]. However, in the present study, the ZnO seed template was eliminated or dissolved during the hydrothermal reaction to form AuNPs directly on the substrate (sacrificial process). Without the ZnO seed template, the AuNP formation cannot be tuned. The ZnO seed template dissolves into the hydrothermal reactive bath during the reaction due to the competitive reaction between Au and ZnO formations, which is attributed to the difference in free Gibbs energy (ΔG; $\Delta G_{Au} = -421.59$ kJ/mol and $\Delta G_{ZnO} = 16.08$ kJ/mol) [22,23]. This preceding approach has the advantages of low-cost setup, uniform AuNP distribution, and adjustable AuNP size and allows large-area fabrication. The effect of the annealing temperature of the seeded ZnO template on AuNP properties after the reaction is explained accordingly. The AuNP formation mechanism using this approach is proposed. The memory properties of PMMSQ-embedded AuNPs are explained in detail.

METHODS

A n-type (100) silicon wafer was cut into small pieces (dimension of 1 × 1 cm) and used as substrates. The silicon substrates were cleaned using a standard RCA cleaning process to remove the organic and inorganic contaminants from the surface. A polymethylsilsequioxane (PMSSQ) layer was deposited at 2,000 rpm for 100 s on the cleaned substrates to achieve 350-nm dielectric layers. Thereafter, the samples were cured in an oven at 160°C for 1 h (Figure 1a). A 200-nm-thick ZnO thin film was then deposited on each substrate using radio-frequency magnetron sputtering at 200 W (Figure 1b). The samples were annealed at different temperatures: 200°C, 300°C, and 400°C, with a ramp rate and soaking time of 5°C/min and 10 min, respectively, to observe the seed layer effects on AuNP formation (Figure 1c). Thereafter, the AuNPs were grown on the ZnO seed template using a sacrificial low-temperature hydrothermal approach. The ZnO-seeded samples were subjected to a hydrothermal reaction in a preheated oven at 80°C for 4 h. The hydrothermal bath contained 0.1 M zinc nitrate tetrahydrate ($Zn(NO_3)_2 \cdot 4H_2O$), 0.1 M hexamethylenetetramine ($C_6H_{12}N_4$), 0.01

M gold (III) chloride trihydrate ($AuCl_4.3H_2O$), and 10 mL acetic acid. After hydrothermal reaction, the samples were removed, rinsed with deionized water, and then dried. The thin ZnO layer was eliminated due to sacrificial growth (Figure 1d).

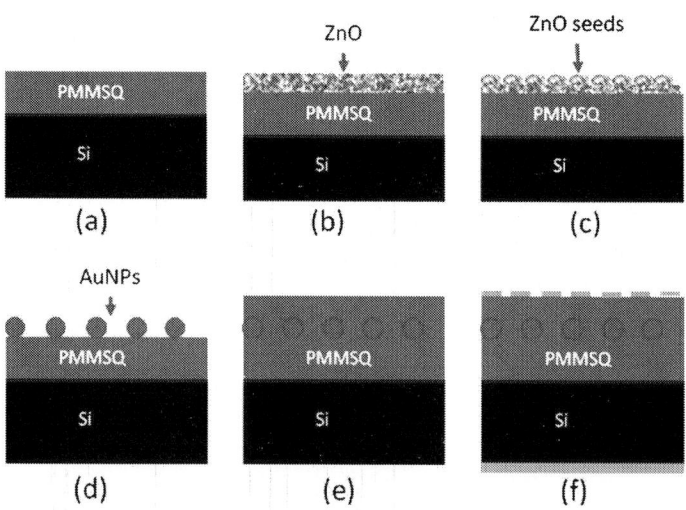

Figure 1: Process flow for sacrificial templated-growth hydrothermal reaction of AuNPs embedded in the PMMSQ memory device. (a) PMMSQ/n-Si, (b) deposited ZnO layer, (c) thermal oxidation of ZnO layer to form ZnO seeds, (d) AuNPs formed on PMMSQ/n-Si, (e) another PMMSQ layer was deposited on the AuNPs, and (f) desired memory device structure with Al as top and bottom electrodes.

A second PMSSQ layer was then spin-coated on top of AuNPs and cured at 160°C for 1 h (Figure 1e). Aluminum (Al) top (1 × 1 mm) and back contacts were deposited using thermal evaporation (Figure 1f).

The surface morphology of the samples was observed using a field emission scanning electron microscope (FESEM; Zeiss Supra™ 35VP, Carl Zeiss AG, Oberkochen, Germany). The presence of chemical elements was analyzed using an energy dispersive X-ray spectrometer. Phase presence was analyzed using the Bruker D8 X-ray diffractometer (XRD; Bruker AXS GmbH, Karlsruhe, Germany). The memory characteristics of the samples were determined using the Keithley Model 4200-SCS semiconductor characterization system (Keithley Instruments Inc., Cleveland, OH, USA).

RESULTS AND DISCUSSION

SEM images of the ZnO seeds with varying annealing temperatures are shown in Figure 2. Before annealing, more large grains were observed (Figure 2a). After annealing, the grain sizes became more uniform with lesser large grains present. The observation proved that the atoms rearranged to form a more stable microstructure during the annealing process. The size of the grains became smaller until annealing at 300°C (Figure 2c). Further increasing the annealing temperature to 400°C caused more large grain presence due to the diffusion of ions (Figure 2d). A similar result was reported by Chu et al. [24]. However, for the hydrothermal growth using a seeded template, the seeds must not be too dense to provide a space for growth during the hydrothermal reaction.

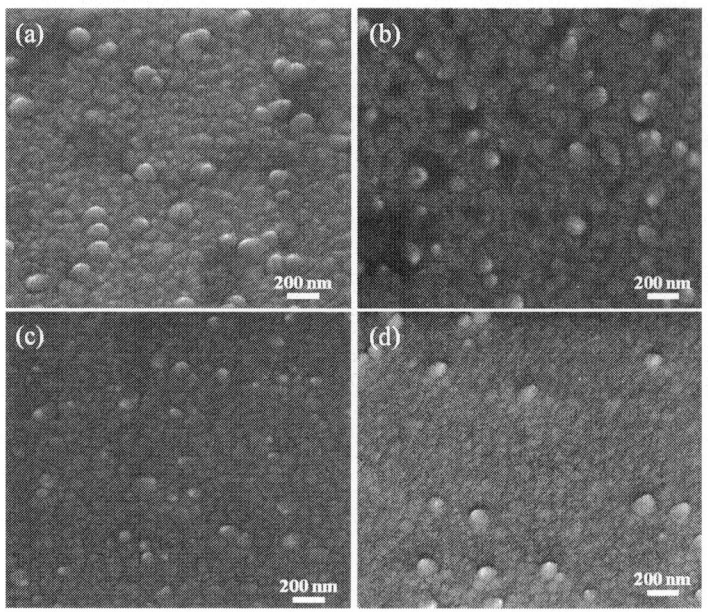

Figure 2: SEM images of the ZnO seed template annealed at varying temperatures. (a) Non-annealed, (b) 200°C, (c) 300°C, and (d) 400°C.

The average seed size of each sample was calculated from SEM images using the ImageJ software (NIH, Bethesda, MD, USA). The ZnO seed sizes of the non-annealed sample and those annealed at 200°C,

300°C, and 400°C were 110, 113, 109, and 104 nm, respectively. The ZnO seeds became smaller with increasing annealing temperature. At high temperature, ZnO atoms vibrate strongly at their lattice positions and exchange energy with neighboring atoms when sufficient thermal energy is supplied during the annealing process. Thus, atoms diffuse to achieve the lowest strain energy [25], forming smaller ZnO seed surface grains (Figure 2).

Figure 3a shows the XRD patterns of the ZnO seed layer template annealed at different temperatures, with a preferential growth along (002) at 34.4° in all samples. The (002) peak intensities increased with increasing annealing temperature, indicating a crystallinity improvement in the ZnO seeds. This crystallinity improvement is due to the sufficient thermal energy of the annealing process, which caused Zn and O atoms to rearrange in a proper site. Thus, the ZnO seed layer template crystallinity is improved [26]. Moreover, the (002) peak positions shifted to a higher angle with increased annealing temperature due to compressive stress relief. During the annealing process, the ZnO film atoms gained energy to rearrange; thus, the ZnO seed film achieved relaxation[24,27,28]. Meanwhile, ZnO peaks disappear; however, a dominant peak of (111) at 38.3°C appeared which corresponds to the AuNP presence. Theoretically, for a face-centered cubic metal Au, the surface energies of the low-index crystallographic facets usually increase in the order of {111} < {100} < {110} due to an increase in interatomic distance. The {110} plane is the most favorable plane for the Au atom deposition because {110} has the highest facet surface energy. Therefore, {110} planes have the highest rate of dissolution and recrystallization of Au atoms, eventually resulting in {110} facet disappearance as the Au particles grow, thereby making the synthesis of Au nanoparticle with {110} facets difficult. Therefore, Au prefers to grow along the {111} facets because the {111} facets have the lowest surface energy, which is more stable, as shown in Figure 3b[29,30]. The ZnO peaks for all samples were unobserved, confirming the dissolution of the ZnO seed template during the hydrothermal reaction. This dissolution is attributed to the competitive growth between Au and ZnO. Furthermore, ZnO erodes in an acidic solution [31]. Therefore, in the present study, the pH of the hydrothermal precursor was similar for all samples at approximately 2.7. Figure 3c shows the cross-sectional structure of the fabricated metal-insulator-semiconductor (MIS) memory device annealed at

400°C. The SEM micrograph showed that AuNPs were embedded in between the two PMSSQ layers.

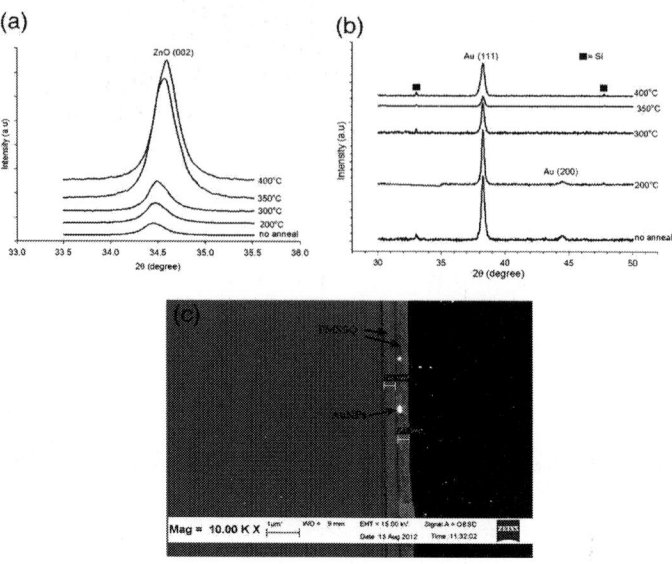

Figure 3: ZnO seed layer template, AuNP growth, and MIS memory device. XRD spectra of the (a) sputtered ZnO seed layer template annealed at various temperatures. (b) Corresponding AuNP growth using the ZnO seed template annealed at various temperatures. (c) Cross-sectional view of the MIS memory device for the sample annealed at 400°C.

Figure 4 shows the morphology of AuNPs grown on the ZnO-seeded substrates annealed at various temperatures after the hydrothermal reaction. The AuNPs for the sample grown on the non-annealed template were spherical and non-uniformly distributed. Furthermore, the AuNPs tend to agglomerate. Fewer large-sized AuNPs were observed for the sample grown on the template annealed at 200°C. Few AuNPs were grown isolated, but some agglomeration was observed. A greater amount of AuNPs with a smaller diameter than the first two samples was obtained for the sample grown on the ZnO template annealed at 300°C, indicating that the AuNP area density was high. More AuNPs were isolated compared with the first two samples. A higher AuNP amount for the samples grown on a template annealed at 400°C was produced, and AuNPs were closer to each other, with some attached to one another.

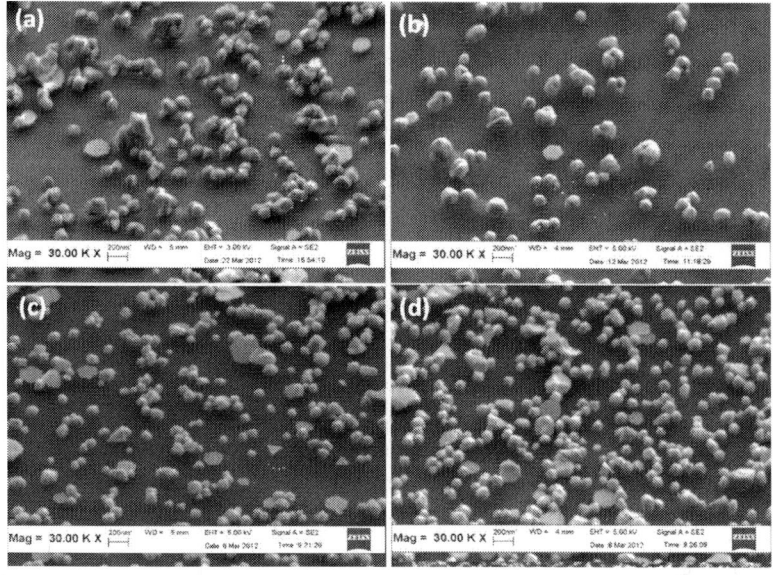

Figure 4: SEM images of AuNPs formed using the ZnO sacrificial template annealed at various temperatures. (a) Non-annealed, (b) 200°C, (c) 300°C, and (d) 400°C.

The AuNP size and distribution were affected by the ZnO seed size. Figure 2 shows that the ZnO seed size decreased with increasing annealing temperature. Thus, more sites were available for AuNP nucleation. With the same Au ion amount, small Au nanoparticles can be formed. The distance between the AuNPs for large seeds was far. Some large AuNPs were also observed due to AuNP interdiffusion.

The average AuNP diameters for the samples grown at different annealing temperatures of the ZnO template were calculated using the ImageJ software. The average AuNP diameters for the non-annealed samples and those annealed at 200°C, 300°C, and 400°C were 113, 120, 90, and 70 nm, respectively. The AuNP number per unit area (area density) was calculated. The area densities of the non-annealed sample and samples annealed at 200°C, 300°C, and 400°C were 2.6×10^{13}, 8.9×10^{12}, 2.7×10^{13}, and 4.5×10^{13} m^{-2}, respectively.

Theoretically, a template determines the size, shape, and distribution of nanoparticles [32]. This basic principle was applicable in this study, wherein the AuNP size and distribution were confined by the template. AuNPs nucleate on a high surface energy site, which is the ZnO grain boundary. The grain boundary area between the ZnO seeds increased and lead to the formation of larger AuNPs when the ZnO seed size increased. The AuNPs grown were larger because of the larger ZnO seeds. Therefore, low AuNP area density was observed.

Formation Mechanism of Au Nanoparticles

During the hydrothermal reaction, the ZnO seed template dissolved into the hydrothermal reactive bath due to the competitive growth between Au and ZnO. There are two systems involved in the hydrothermal bath, as shown in Equations 1 and 2.

$$AuCl_4^- + 3e^- \leftrightarrow Au + 4Cl^-$$

(1)

$$Zn(OH)_2 \leftrightarrow ZnO + H_2O$$

(2)

The ΔG of Au and ZnO formations were −421.59 and −16.08 kJ/mol, respectively [21, 22]. Based on the difference in ΔG values, AuNP formation was more favorable; thus, ZnO formation was suppressed. Furthermore, the precursor solution was acidic; ZnO erodes in an acidic condition [31]. ZnO is amphoteric and dissolves in acids to form salts that contain hydrated zinc (II) cation [33]. Consequently, $Zn(OH)_2$ dissociated back into OH^- and Zn^{2+} to neutralize the solution by reacting with H^+ ions in the precursor solution, as shown in Equations 3 and 4. Therefore, pH of the solution after the hydrothermal reaction increased.

$$Zn(OH)_2 \leftrightarrow Zn^{2+} + OH^-$$

(3)

$$H^+ + OH^{--} \leftrightarrow H_2O$$

(4)

During the hydrothermal reaction, Au ions nucleated between the ZnO seed layer because the surface energy is higher at the grain boundaries. Au ions from the hydrothermal solution diffused on the AuNP nuclei and formed larger particles. At the same time, the ZnO seed layer acted as a sacrificial template for AuNP growth because ZnO dissolved into the solution after the AuNP nuclei started to grow. The formation mechanism is illustrated in Figure 5.

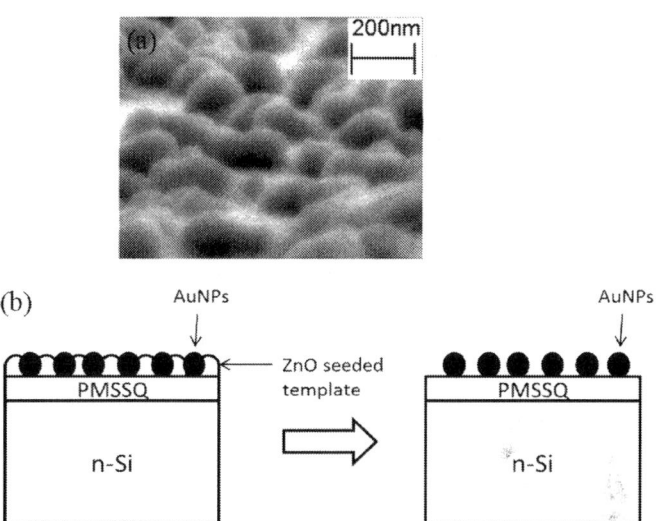

Figure 5: SEM image and schematic diagram. (a) SEM image of Au nanoparticles grown using the ZnO template. (b) Schematic diagram of Au nanoparticle formation between the ZnO seed layer and the ZnO-seeded template dissolution in the acidic hydrothermal reactive solution.

Electrical Properties of PMMSQ-embedded AuNPs

About 350 nm of PMSSQ was spin-coated on the AuNPs to investigate the electrical properties of the produced AuNPs. The top and bottom contacts were prepared using Al. The current–voltage (*IV*) characteristics of PMMSQ-embedded AuNPs with varying annealing temperatures of the ZnO template are shown in Figure 6. The *IV* characteristics were obtained by applying positive voltage on the Al top contact with respect to the Al bottom contact. For all samples, the positive voltage was swept from 0 to 10 V and vice versa. Prior to further investigation of the AuNP effects in the MIS memory structure, the negligible charge-trapping capability without AuNPs between PMSSQ through the capacitance-voltage (*C-V*) hysteresis windows has been reported previously [34]. During the reverse sweep, the current increased in all memory devices due to a trapped electron at the AuNP sites [1].

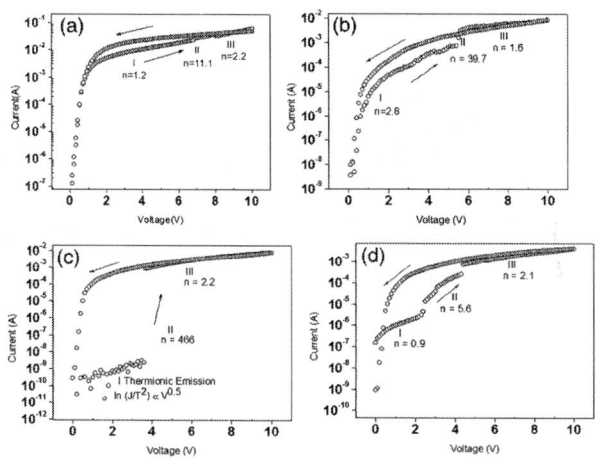

Figure 6: *I-V* characteristics. (a) Non-annealed sample, (b) annealed at 200°C sample, (c) annealed at 300°C sample, and (d) annealed at 400°C sample as MIS memory devices.

The *I-V* characteristics in Figure 6 were marked as regions I, II, and III to explain the possible memory device transport mechanisms. The *I-V* relationship can be expressed as $I \propto V^n$, and the fitted slope of a log-log plot determines the *n* value of $\log I \propto n \log V$. During the forward sweep, an abrupt current increment can be observed at the threshold voltage, V_t, in region II. At V_t, electrons from the n-type silicon substrate tunnel into the AuNP acceptor-like centers through the PMSSQ layer. In the low-voltage region I ($V < V_t$), ohmic conduction, space-charge-limited current (SCLC), and thermionic emission (TE) are the dominant conduction mechanisms, where electrons are being injected from the n-type Si into AuNPs. In ohmic conduction, $n = 1.2$ and $n = 0.9$ for the memory devices that were non-annealed and annealed at 400°C, respectively. In SCLC, $n = 2.8$ for the memory device annealed at 200°C, whereas the log-log plot for the memory device annealed at 300°C does not fit the $\log I \propto n \log V$ relationship but fits the $\log e(J/T2) \propto V0.5$, where J and T are the current density and temperature of the system, indicating that the TE mechanism was obeyed. As the voltage increased to V_t, the transport mechanism switched to the trapped charge-limited current for all memory devices because $n >> 2$ as marked in region II. In this region, the trap sites due to the presence of AuNPs started to be filled by electrons, and an abrupt current increase was observed. After all traps in the AuNPs were filled, the transportation mechanism switched to trap-free (an ideal SCLC transport mechanism has $n = 2$) SCLC for all cases. During the reverse sweep, the current flow through the device remained high because all the traps were filled and followed the SCLC transport mechanism. The electrons can be stored and retained, thereby proving the existence of the memory effect. An energy-band diagram illustrates the electron injection from Si to PMSSQ and trapped in AuNPs, as described in Figure 7.

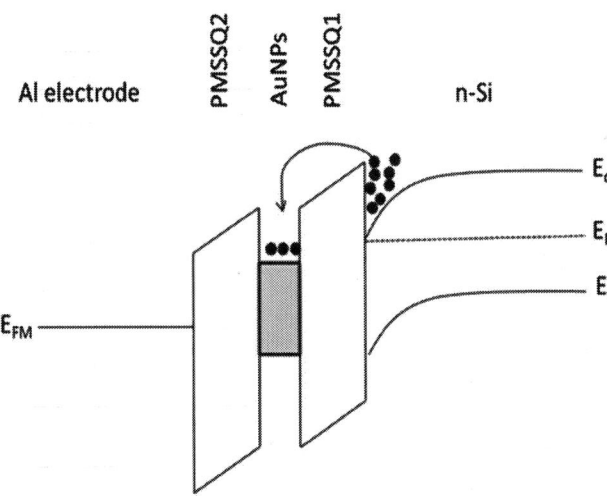

Figure 7: Energy-band diagram. The diagram describes the electron flows and trap observed during the positive bias on the Al top electrode with respect to the Al bottom contact.

The sample prepared using AuNPs grown on the non-annealed template showed an abrupt increase at 6.8 V, which was the highest V_t observed among all samples, with 0.8 orders of increment in current magnitude. This preceding aspect was due to the low area density of the AuNPs formed. For the sample grown on the template annealed at 200°C, 0.7 orders of current magnitude increment was observed with lower V_t, 5.6 V. The low AuNP area density was compensated by the larger and isolated AuNPs that stored more charges to obtain a similar increment with the first sample, which was supported by the findings on the effect of size on the memory window [35, 36]. The largest hysteresis was obtained for the sample prepared on the ZnO template annealed at 300°C, with 6 orders of current magnitude increment at 3.6 V, indicating that uniform AuNPs observed in combination with their size and area density exhibited excellent memory effects. For the sample grown on the template annealed at 400°C, 2 orders of current magnitude increment was observed at 2.4 V, which is attributed to the high area density of isolated AuNPs. Based on the *IV* characteristics, the samples prepared using templates annealed at 300°C and 400°C showed larger current increment. Both samples showed good distribution of isolated AuNPs when related back to the microstructure. The sample grown on the template annealed at 300°C produced better *IV* responses between an erased and programmed device. *IV* plot is important as it allows the understanding of the charge transport mechanism when the device is erased or programmed.

The *C-V* characteristics of the PMMSQ-embedded AuNPs using the ZnO template annealed at varying temperatures are shown in Figure 8. All samples were swept from the negative to the positive voltage and then back to the negative voltage. The hysteresis (flat-band voltage shift, ΔV_{FB}) window for the sample prepared on the non-annealed template was 3.7 V, which was the largest hysteresis. The sample prepared on the template annealed at 200°C exhibited the smallest hysteresis of 1.2 V. Meanwhile, a large hysteresis of 3.6 V was observed in the sample grown on the template annealed at 300°C. No complete hysteresis window was observed for the sample grown on the template annealed at 400°C because the AuNPs formed were too dense and led to the lateral flow of charges. Hence, the charge per Au nanoparticle was not calculated.

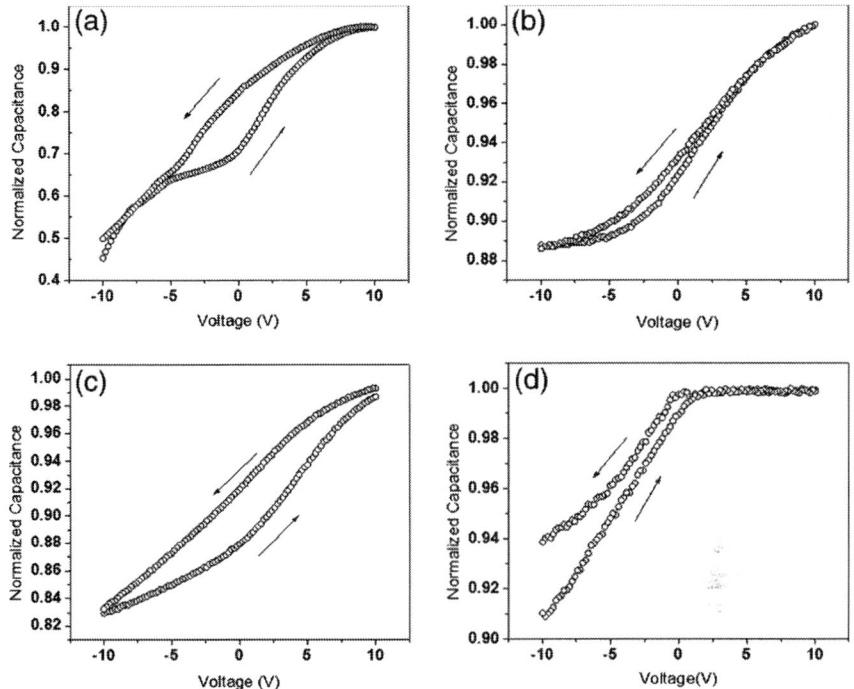

Figure 8: C-V characteristics of PMMSQ-embedded AuNPs. (a) Non-annealed samples and those (b) annealed at 200°C, (c) 300°C, and (d) 400°C.

The following illustrates the calculation of charge storage capacity per AuNP. Flat-band voltage shift, ΔV_{FB}, was calculated for a single electron ($n = 1$) confined in AuNPs using Equation 5:

$$\Delta V_{FB} = \frac{nqd_{\text{nanoparticle}}}{\varepsilon_{\text{PMSSQ}}}\left(t_{\text{gate}} + \frac{1}{2}D_{\text{nanoparticle}}\right)$$

(5)

where n is the number of charges per single Au nanoparticle, q is the electron charge magnitude, $d_{\text{nanoparticle}}$ is the area density, t_{gate} is the thickness of the control gate (PMSSQ), $D_{\text{nanoparticle}}$ is the AuNP diameter, and $\varepsilon_{\text{PMSSQ}}$ is the PMSSQ dielectric constant.

Through the observed flat-band voltage shift, $\Delta V_{FBO,}$ from the C-V curve, the amount of charges stored per single Au nanoparticle can be calculated as follows:

$$\text{Stored charge density, } nd_{\text{nanoparticle}}$$
$$= \frac{\Delta V_{\text{FB}} \varepsilon_{\text{PMSSQ}}}{q \left(t_{\text{gate}} + \frac{1}{2} D_{\text{nanoparticle}} \right)}$$

$$(6)$$

Therefore, the number of stored charge per single Au nanoparticle was obtained through the stored charge density in Equation 6, divided by area density, which is represented as

$$\frac{nd_{\text{nanoparticle}}}{d_{\text{nanoparticle}}} \, .$$

The data obtained through the *CV* characteristics of each sample are summarized in Table 1. Large hysteresis was found for the sample grown on the non-annealed template and at 300°C, 3.7 and 3.6 V, respectively. From the microstructure observation, the latter sample had a higher AuNP area density but had a smaller diameter. However, the hysteresis window size was slightly smaller compared with the sample grown on the non-annealed template. Therefore, a larger Au nanoparticle size is probably better for charge storage, which agrees with Tseng and Tao [35]. The amount of charge stored per Au nanoparticle grown on the template annealed at 300°C was 54, which is the highest compared with the samples grown on the non-annealed and at 200°C (43 and 37, respectively). The results proved that the presence of isolated AuNPs is important to obtain optimum memory properties.

Table 1: Summary of **Δ** *V* FBand amount of electrons stored per AuNPs

	Annealing temperature		
	Non-annealed	200°C	300°C

Hysteresis or flat-band voltage shift, VFB(V)	3.7	1.2	3.6
Amount of electrons stored per Au nanoparticle	43	37	54

The data shown are for the non-annealed samples and those annealed at 200°C and 300°C.

Goh *et al.*

Goh *et al. Nanoscale Research Letters* 2012 **7**:563, doi:10.1186/1556-276X-7-563

CONCLUSIONS

AuNPs were successfully grown directly on PMSSQ using a sacrificial templated growth hydrothermal reaction at a low temperature. The AuNP size and distribution were dependent on the annealing temperature of the ZnO sacrificial template because AuNPs were nucleated and grown on ZnO seed grain boundaries. The AuNP size decreased, whereas the area density increased, with increasing ZnO annealing temperature. Optimum memory properties of the PMMSQ-embedded AuNPs were obtained for AuNPs grown on the ZnO seed template annealed at 300°C, with an estimated 54 electrons per Au nanoparticle and excellent *I-V* responses between an erased and programmed device.

AUTHORS' CONTRIBUTIONS

LPG and NSR performed the experimental works and drafted the manuscript. KAR designed the experimental works as well as the writing of the manuscript. CKY and KCA participated in designing and advising the electrical properties analysis. PCO performed the electrical properties analysis and data interpretation. All authors read and approved the final manuscript.

ACKNOWLEDGMENTS

The authors gratefully acknowledge the technical support from the School of Materials and Mineral Resources Engineering, the Institute for Research in Molecular Medicine, and the Nano-Optoelectronics Research and Technology Lab, USM. This research was jointly supported by the USM Research University grants (1001/PSKBP/8630019 and 1001/PBahan/814135), LRGS UTP, and ERGS.

REFERENCES

1. Park B, Im K-J, Cho K, Kim S: Electrical characteristics of gold nanoparticle-embedded MIS capacitors with parylene gate dielectric. *Org Electron* 2008, 9:878.

2. William S, Mabrook MF, Taylor DM: Floating-gate memory based on an organic metal-insulator-semiconductor capacitor. *Appl Phys Lett* 2009, 95:093309.

3. Scott JC: Is there an immortal memory? *Science* 2004, 304:62.

4. Prime D, Paul S: Overview of organic memory devices. *Phil Trans R Soc A* 2009, 367:4141.

5. Yu H, Chen Y, Huang C, Su Y: Investigation of nonvolatile memory effect of organic thin-film transistors with triple dielectric layers. *Appl Phys Express* 2012, 5:034101.

6. Müller K, Henkel K, Paloumpa I, Schmeißer D: Organic field effect transistors with ferroelectric hysteresis. *Thin Solid Films* 2007, 515:7683.

7. Fujisaki S, Ishiwara H, Fukisaki Y: Low-voltage operation of ferroelectric poly(vinylidene fluoridetrifluoroethylene) copolymer capacitors and metal-ferroelectric-insulator-semiconductor diodes. *Appl Phys Lett* 2007, 90:162902.

8. Cagin E, Chen DY, Siddiqui JJ, Phillips JD: Hysteretic metal–ferroelectric– semiconductor capacitors based on PZT/ZnO heterostructures. *J Phys D: Appl Phys* 2007, 40:2430.

9. Naber RCG, Tanase C, Blom PWM, Gelinck GH, Marsman AW, Touwslager FJ, Setayesh S, Leeuw DMD: High-performance

solution-processed polymer ferroelectric field-effect transistors. *Nat Mater* 2005, 4(3):243-248.

10. Lee J: Recent progress in gold nanoparticle-based non-volatile memory devices. *Gold Bull* 2010, 43:189.

11. Tsoukalas D: Metallic nanoparticles for application in electronic non-volatile memories. *Int J Nanotechnol* 2009, 6:35.

12. Lee C, Gorur-Seetharam A, Kan EC: Operational and reliability comparison of discrete-storage nonvolatile memories: advantages of single- and double-layer metal nanocrystals. *IEDM Tech Dig* 2003, 3:557.

13. Ipe BI, Thomas KG, Barazzouk S, Hotchandani S, Kamat PV: Photoinduced charge separation in a fluorophore-gold nanoassembly. *J Phys Chem B* 2002, 106:18.

14. Sargentis C, Giannakopoulos K, Travlos A, Tsamakis D: Fabrication and electrical characterization of a MOS memory device containing self-assembled metallic nanoparticles. *Physica E* 2007, 38:85.

15. Han ST, Zhou Y, Xu ZX, Roy VAL, Hung TF: Nanoparticle size dependent threshold voltage shifts in organic memory transistors. *J Mater Chem* 2011, 21:14575.

16. Prakash A, Ouyang J, Lin J, Yang Y: Polymer memory device based on conjugated polymer and gold nanoparticles. *J Appl Phys* 2006, 100:054309.

17. Paul S, Pearson C, Molloy A, Cousins MA, Green M, Kolliopoulou S, Dimitrakis P, Normand P, Tsoukalas D, Petty MC: Langmuir-Blodgett film deposition of metallic nanoparticles and their application to electronic memory structures. *Nano Lett* 2003, 3:533.

18. Kim H, Jung S, Kim B, Yoon T, Kim Y, Lee H: Characterization of charging effect of citrate-capped Au nanoparticle pentacene device. *J Ind Eng Chem* 2010, 16:848.

19. Gupta RK, Krishnamoorthy S, Kusuma DY, Lee PS, Srinivasan MP: Enhancing charge-storage capacity of non-volatile memory devices using template-directed assembly of gold nanoparticles. *Nanoscale* 2012, 4:2296.

20. Tan WK, Razak KA, Ibrahim K, Lockman Z: Formation of ZnO nanorod arrays on polytetraflouroethylene (PTFE) via a seeded

growth low temperature hydrothermal reaction. *J Alloys Compd* 2011, 509:820.

21.   Lockman Z, Pet Fong Y, Wai Kian T, Ibrahim K, Razak KA: Formation of self-aligned ZnO nanorods in aqueous solution. *J Alloys Compd* 2010, 493:699.

22.   Davies A, Staveley LAK: The thermodynamics of the stable modification of zinc hydroxide, and the standard entropy of the aqueous zinc ion. *J Chem Thermodyn* 1972, 4:267.

23.   Marsden JO, Lain CI: *The Chemistry of Gold Extraction*. Colorado: Society for Mining, Metallurgy and Exploration Inc; 2006.

24.   Chu SY, Water W, Liaw JT: Influence of postdeposition annealing on the properties of ZnO films prepared by RF magnetron sputtering. *J Eur Ceram Soc* 2003, 23:1593.

25.   Mittemeijer EJ: *Fundamentals of Materials Science: The Microstructure-Property Relationship Using Metals as Model Systems*. New York: Springer; 2011.

26.   Liu SY, Chen T, Wan J, Ru GP, Li BZ, Qu XP: The effect of pre-annealing of sputtered ZnO seed layers on growth of ZnO nanorods through a hydrothermal method. *Appl Phys A* 2009, 94:775.

27.   Karamdel J, Dee CF, Majlis BY: Effects of annealing conditions on the surface morphology and crystallinity of sputtered ZnO nano films. *Sains Malaysiana* 2011, 40:209.

28.   Sun X, Fu Z, Wu Z: Fractal processing of AFM images of rough ZnO films. *Mater Charact* 2002, 48:169.

29.   Personick ML, Langille MR, Zhang J, Harris N, Schatz GC, Mirkin CA: Synthesis and isolation of {110}-faceted gold bipyramids and rhombic dodecahedra. *J Am Chem Soc* 2011, 133:6170.

30.   Bakshi MS, Sachar S, Kaur G, Bhandari P, Biesinger MC, Possmayer F, Petersen NO:Dependence of crystal growth of gold nanoparticles on the capping behavior of surfactant at ambient conditions. *Cryst Growth Des* 2008, 8:1713.

31.   Baruah S, Dutta J: pH-dependent growth of zinc oxide nanorods. *J Cryst Growth* 2009, 311:2549.

32.   Hulteen JC, Martin CR: A general template-based method for the preparation of nanomaterials. *J Mater Chem* 1997, 7:1075.

33. Degen A, Kosec M: Effect of pH and impurities on the surface charge of zinc oxide in aqueous solution. *J Eur Ceram Soc* 2000, 20:667.

34. Ahmad Z, Ooi PC, Aw KC, Sayyad MH: Electrical characteristics of poly(methylsilsesquioxane) thin films for non-volatile memory. *Solid State Commun* 2011, 151:297.

35. Tseng C, Tao Y: Electric bistability in pentacene film-based transistor embedding gold nanoparticles. *J Am Chem Soc* 2009, 131:12441.

36. Houili H, Tutis E, Izquierdo R: Modeling nanoparticle embedded organic memory devices. *Org Electron* 2010, 11:514.

Raw Materials Synthesis from Heavy Metal Industry Effluents with Bioremediation and Phytomining: A Biomimetic Resource Management Approach

Salmah B. Karman[1,2], S. Zaleha M. Diah[1], and Ille C. Gebeshuber[1,3]

[1]Institute of Microengineering and Nanoelectronics, Universiti Kebangsaan Malaysia, 43600 Bangi, Malaysia

[2]Department of Biomedical Engineering, Faculty of Engineering, University of Malaya, 50603 Kuala Lumpur, Malaysia

[3]Institute of Applied Physics, Vienna University of Technology, Wiedner Hauptstraße 8-10/134, 1040 Vienna, Austria

ABSTRACT

Heavy metal wastewater poses a threat to human life and causes significant environmental problems. Bioremediation provides a sustainable waste management technique that uses organisms to remove heavy metals from contaminated water through a variety of different processes. Biosorption involves the use of biomass, such as plant extracts and microorganisms (bacteria, fungi, algae, yeast), and represents a low-cost and environmentally friendly method of bioremediation and resource management. Biosorption-based biosynthesis is proposed as a means of removing heavy metals from wastewaters and soils as it aids the development of heavy metal nanoparticles that may have an application within the technology industry. Phytomining provides a further green method of managing the metal content of wastewater. These approaches represent a viable means of removing toxic chemicals from the effluent produced during the process of manufacturing, and the bioremediation process, furthermore, has the potential to save metal resources from depletion. Biomimetic resource management comprises bioremediation, biosorption, biosynthesis, phytomining, and further methods that provide innovative ways of interpreting waste and pollutants as raw materials for research and industry, inspired by materials, structures, and processes in living nature.

INTRODUCTION

Heavy metals are hazardous and dangerous, especially when introduced into the environment via pollution. However, they do have a valuable role to play; for example, heavy metal nanoparticles are used in various nanoscience and nanotechnology applications. Biomimetic resource management refers to a way of dealing with resources that is inspired by living nature regarding materials, structures, and processes. Such an approach offers innovative new ways to deal with heavy metal-loaded waste effluents and provides raw materials for industry. Plants and microorganisms are used to redefine "waste" to "revenue" for new industries [1]. Heavy metal-loaded effluents from industry could be the base material for metallic nanoparticles used in nanoscience and nanotechnology. Plants (such as the sunflower plant) and microorganisms (such as bacteria, fungi, algae, and yeast) can be used to accumulate these heavy metals and to safely remove the pollutants from the water and the soil. The first goal regarding heavy metal effluents should be their prevention. However, in cases where they cannot (yet) be completely prevented, biomimetics can come into the game: we could learn from living nature how to deal with such effluents not by treating them as waste but by treating them as resource (waste-to-wealth concept). Metallic nanoparticles are currently used in important nanotechnology research areas and are also heavily used in applications. They are important (and acceptable) in our current early phases of nanotechnology research and development, where we need to understand the basics. Future resource management might increasingly realize the paramount biomimetic principle of "shape rather than material" to achieve functionalities that are currently fulfilled by unsustainable metal- and plastic-based resources by benign materials that would allow for sustainable engineering. Apart from biomineralized structures and specific biomolecules, where the chemistry (as opposed to the physics) of the metal is necessary for the function (such as in hemoglobin or chlorophyll), living nature rarely uses metals. In most cases, elaborated structures from hierarchically composed metal free materials yield the functionality that we, with our current conventional engineering, mainly achieve with the use of many different materials, including metals.

The main purpose of this paper is to explore the potential of using heavy metals that are extracted from the environment with the help of organisms as a resource for metallic nanoparticles. This has a number of potential benefits. The particles are removed from the environment and research, development, and industry are provided with a novel source of nanoparticles and their constituents. Since some nanoparticles can be derived directly from natural sources, such as plants and microorganisms, they can be accessed at a lower cost and in a more biofriendly manner than those fabricated via conventional production mechanisms. Another advantage of extracting heavy metals from the environment is that the natural nanoparticles accessed via this method exhibit reproducible shapes and sizes, whereas such shape and size uniformity remains a challenge in man-made nanoparticles. A variety of physical and chemical procedures are used for the synthesis of metallic nanoparticles, with "bottom-up" and "top-down" approaches as two broad categories.

This review predominantly concentrates on the management and recovery of heavy metals from industrial waste-water through the application of a biosorption-desorption process. The production of metal nanoparticles from heavy metal industrial effluent is also discussed. A particular focus will be placed on the bioremediation process, since this method holds great promise as a potential technique of biosynthesizing metal nanoparticles from polluted heavy metal industrial effluent. It is envisaged that developments in this area provide a means of removing toxic effluent from wastewater and/or soil and also represent a viable and sustainable technique for obtaining nanoparticles for high technology applications in a manner that is environmentally friendly and cost effective.

THE BIOREMEDIATION PROCESS AS AN APPROACH FOR DESIGNING MATERIALS FOR ENVIRONMENTAL APPLICATIONS

In order to ensure that industrial practices are sustainable on a long-term basis, it is critical that we develop an awareness of the environment and the ecological effects of manufacturing techniques that produce toxic metals. Contaminated soil in the environment as a result of manufacturing practices is common throughout the world. Numerous countries face issues as a result of contaminated wastewater and have taken action to raise awareness via relevant policies and the development of technologies [2]. However, many of the physiochemical processes that are traditionally employed in the remediation of soils and polluted sites are expensive and do not permanently alleviate hazardous pollution [3]. Bioremediation represents one method of using biological systems, such as microorganisms or microbial-like bacteria, fungi, and other agents, to clean up and degrade organic and inorganic pollutants [4]. Bioremediation is a general concept that encompasses all the processes and actions required to biotransform an environment in which contaminants exist back into its original pristine condition. Various factors are involved in the bioremediation process and this process uses various agents, such as bacteria, yeast, fungi, algae, and higher plants, as major tools to treat oil spills and remove heavy metals from the environment [5]. A number of sophisticated technologies for the remediation of polluted sites are currently in use.

The Bioremediation Process: Current Situation

Effluents produced by the metal industry are often found as contaminants in water sources, rivers, seas, and soils. Fu and Wang [6] published a review on the use of bioremediation to remove heavy metal ions from wastewaters; remediation technologies used to treat heavy metal contaminated groundwater were summarized by Hashim et al. [7]. Heavy metals are defined as metals that have an atomic weight between 63.5 and 200.6 and a specific gravity greater than 5 [7]. They include Zinc, Copper, Nickel, Mercury, Cadmium, Lead, Chromium Arsenic, Silver, Platinum, and Gold [6]. Some heavy metals are toxic and poisonous, especially to humans, while noble metals are very valuable and can be used in high-tech nanotechnology applications or in the production of high-value goods.

The management of toxic heavy metals that are introduced into the environment via industrial wastewater is very important, as toxic wastewaters pose serious threats to the environment and to human health [8]. The technologies that are currently used in the commercial remediation of heavy metal effluent rely on immobilizing the heavy metal by leachability [9]. Those techniques include chemical fixation, chemical alteration/complexation, stabilisation, capping, soil washing, and ferric iron remediation stabilisation. Additional methods that are currently being used to remediate heavy metal contamination are chemical precipitation, ion exchange, adsorption, membrane filtration, electrochemical treatment technologies, floatation, coagulation, and flocculation [6]. Each method has its own advantages and limitations. The chemical precipitation, ion exchange, and membrane filtration methods are widely used for wastewater remediation with high efficiency. In the chemical precipitation method, the heavy metal ion is altered by changing it from a soluble to an insoluble substance so that it can later be removed via a process of flocculation and sedimentation. Through a combination of chemical precipitation and nanofiltration methods, it is possible to recover heavy metal ions for reuse.

The main limitations of these methods are that they are costly, involve complex processes, and are environmentally unfriendly. For example, the ion exchange method is expensive and the chemical handling process it employs could cause secondary pollution. Even though the use of natural resin, zeolites, is low-cost and environmentally friendly and offers the same performance as synthetic resin, its availability is limited and, at present, its use is limited to the experimental laboratory only.

The adsorption process has long been recognized as a low-cost alternative to the removal of contaminants from wastewater. This flexible, low-cost method works even in wastewater that contains a low concentration of metal effluents. However, the efficiency of this method varies according to the type of adsorbents, since certain sorbents have high selectivity towards heavy metals.

In commercial adsorption-based remediation processes, commercial activated Carbon (AC) is widely used as a high efficiency absorbent [6]. However, the price of AC is increasing. Furthermore, when used as absorbents, Carbon nanotubes, for example, pose a risk to humans when they are discharged into the water. To avoid various limitations, especially the risk to humans, there is a need to identify a means of improving this method so that it is more environmentally friendly and of zero risk to life forms. Furthermore, as a result of the rapid growth of industrialization, which has increased the volume of heavy metal pollution, there is a need to identify a low-cost method of removing contaminants from wastewater. Biology-based remediation processes are attractive because they have lower operation costs than physicochemical processes [10]. Thus, the use of biological materials for absorption holds great promise as they are highly cost effective and environmentally benign. As such, a large amount of research has been conducted into the use of biological materials for the absorption of heavy metals in wastewater with much of this research placing a particular emphasis on the types of biosorbents that are available.

Biosorbents from living cells, dead cells, or biomass are the key components in biosorption technology and, for the last 20 years, various researchers have focused on these materials [10]. As reported by Atkinson et al. [10], the earliest investigation into microbial biomass as a biosorbent was performed in the 1980s, potentially earlier. Current studies on bacterial biosorbents for the bioremediation of heavy metals were reviewed by Dhankhar and Guriyan [11] and Vijayaraghavan and Yun [12]. Studies on marine algae as biosorbents were comprehensively reviewed by He and Chen [13]. Bankar et al. [14] and Viraraghavan and Srinivasan [15], respectively, discussed recent reviews on yeasts and fungal-based biosorption. Studies on biomass from agricultural waste and byproducts for the removal of heavy metals have become a new interest in recent years [16–18].

Removing Heavy Metals

Heavy metals are present in soil, aqueous solution, or streams as the result of a variety of human waste activities, which include intensive agriculture, sludge dumping, metal-rich mine tailings, metal smelting, electroplating, energy conversion, and fuel production [19]. All heavy metals have a toxic effect if they are found in high concentrations in soil and therefore need to be removed or transformed. Microorganisms can be used as cation sorbents for the removal of heavy metal cations from industrial wastewater or for the recovery of metals from their solutions [20]. Dave and Chopda [21] described how a surface modification strategy could enhance the stability and efficiency of iron oxide nanomaterials in the removal of heavy metals for remediation in water. Metal oxide nanoparticles as antimicrobial additives have been the subject of extensive research [22]. Ahluwalia and Goyal [23] described the removal of heavy metals such as Lead, Zinc, Cadmium, Chromium, Copper, and Nickel from wastewater through the use of microbial and plant-derived biomass of Aspergillus niger, Penicilium chrysogenum, Rhizophus nigricans, Ascophyllum nodosum, Sargassum natans, Chrorella fusca, Oscillatoria anguistissima, Basillus firmus, and Streptomyces sp. The ability of algae to remove heavy metals from aqueous solution has been recognized for some decades. Li and coworkers [24] carried out an experiment using the yeasts Zygosaccharomyces rouxii and Saccharomyces cerevisiae in Cadmium removal in a complex food environment. Their results indicated that Z. rouxii had a powerful Cadmium removal ability at low Cadmium concentrations, which mainly depended on the intracellular Cadmium bioaccumulation. The percentage of intracellular Cadmium bioaccumulation of both Z. rouxii and S. cerevisiae decreased as the initial biomass and Cadmium concentrations increased. The metal content of algae can be used to predict the level of metal pollution in a water body [25]. The high accumulation capacity can even be used for the enrichment or recycling of valuable metals. Their relative comparison is generally made using an accumulation factor (AF). The metal accumulation factor (AF) is defined as the ratio of metal concentration in plant cells ($\mu g/g$) and the metal concentration in water ($\mu g/mL$) and it is also known as bioconcentration ratio, concentration ratio, or enrichment ratio. Toxic metals can be transferred to their surroundings or to the wider environment through numerous ways including industrial production processes, incineration emissions, and waste disposal. The majority of deposition of metals in the environment is within soil or sediment. Microorganisms can detoxify metals by valence transformation, extracellular chemical precipitation, or volatilization, through which they enzymatically reduce some metals through metabolic processes [26] (Table1). New technologies to improve the remediation process are continually being developed and adopted and a system known as "pump-and-treat" has been established; however, this is time consuming and inefficient [27].

Table 1: Organisms that remove heavy metal from waste (selection)

Element	Waste from ...	Organisms	Mechanism
Copper [28]	Electronic waste	Leaf extract weed Lantana camara	
Copper [28]	Electronic waste	Fusarium and Pseudomonas	
Arsenic	Water	Lactobacillus acidophilus	Bind and remove
Chromium (IV) [29]	Aqueous solution	Guar gum	Nanozinc oxide biocomposite
Chromium (IV) [30]	Aqueous solution	Seaweed biomass Acanthophora spicifera	Biosorption
Cadmium (Cd)	Food environment	Yeast Zygosaccharomyces rouxii and Saccharomyces cerevisiae	Intracellular cross

Phytoremediation

Phytoremediation technology can be applied as a solution to the major environmental and human hazards caused by contaminated soils and waters [31]. Phytoremediation involves the use of various green plant species to clean up, remove, or detoxify environmental contaminants or to render them harmless [31–33]. Phytoremediation is potentially the best practice for removing pollutants, is very promising as an environmental technology, and can be employed at a lower cost than conventional or alternative methods. Phytoremediation technologies are quite successful in their ability to clean up waste solutions. The phytoremediation approach exploits the ability of various plant species to remove heavy metals from the environment and then accumulate a large amount of toxic metals. The green plant species required to be effective in phytoremediation typically needs to grow rapidly, produce large quantities of biomass, have deep roots and easily harvested shoots, and have the potential to accumulate high concentrations of contaminants in these shoots. The advantages of this method compared to the existing remediation techniques are that they involve minimal site destruction and destabilization, have a low environmental impact, and are aesthetically favorable [19].

One important characteristic of phytoremediation is phytoextraction. Phytoextraction technology involves the reduction of metal concentration in the soil through the cultivation of plants that have a high capacity for metal accumulation in the shoots or that are capable of uptaking the metal from the contaminated soil via the plant root. Phytoextraction technology is a plant-based technology that cleans heavy metal from the soil through the process of hyperaccumulation [34]. This technology is used to achieve better environmental protection with a sustainable metal source and is based on the hyperaccumulation of metals into the whole plants [35]. Two categories of plants can be used for phytoextraction: small plants that have high foliar metal concentration but slow growth rates that do not provide a high annual biomass (such as Thlaspi caerulescens) and high biomass crops which have a large biomass production but take up lower metal concentrations (such as Brassica juncea). Gupta and coworkers [36] used Phaseolus vulgaris var. T55 for the phytoextraction of heavy metals from flash ash (by-product of combustion of coal). The capability of the plants to reduce the amount of heavy metals present in contaminated soils depends on plant biomass production and their metal bioaccumulation factor. The bioaccumulation factor is the ratio of metal concentration present in the shoot tissue of the roots to the one of the soil. It can be determined by the ability and capacity of the roots to take up metals and transfer them to the xylem through the mass flow in xylem by transpiration and their ability to accumulate, store, and detoxify metals while maintaining metabolism, growth, and biomass production [37].

Characterization of Metal Biosorption Mechanisms

The biological interaction between biological cells and metal ions has potential in the production of metal nanoparticles or metal compounds. These processes are very popular in the bioremediation of soil and water sources that have been contaminated by heavy metal effluent [7]. The detailed features of the metal nanoparticles were obtained and the performance of the process depends on the respective type of biological cells and metal ions [12]. The interaction processes that are involved include biosorption, accumulation, and bioreduction [7, 12].

The mechanism for binding metals to the walls of bacterial cells consists of three steps: (i) ion exchange reactions with peptidoglycan and teichoic acid, (ii) precipitation through nucleation reactions, and (iii) complexation with nitrogen and oxygen ligands [38, 39].

The bioaccumulation process is the use of metabolic activity of a living organism to remove heavy metal from a given environment [12, 40].

The biosorption process involves the extracellular passive binding of a metal to a nonliving biomass in an aqueous solution. Two types of approaches commonly used are biosorption without enzyme (a.k.a. protein capping) and biosorption with capping of enzyme. Biosorption has already been investigated for the decontamination of heavy metal pollution solutions [41]. Biosorption technology provides one potential biological approach to cleaning heavy metal industrial wastewater [10]. A number of review papers have discussed the use of biosorption technology in the bioremediation process.

The biosorption process always depends on a number of mechanisms such as complexation, ion exchange, coordination, adsorption, desorption, chelation, and microprecipitation [42, 43]. In general, desorption is one of the most common mechanisms that are employed in the biosorption process [44]. This process is the reverse of sorption (adsorption and absorption). The importance of desorption lies in its potential to actually reuse certain biomass as well as recover sorbents [16, 44]. Through the use of this biosorption-desorption process, low-cost heavy metal nanoparticles could be biologically synthesized during the bioremediation process. As far as the literature reviewed for the purposes of this paper, bioabsorption is typically regarded as a passive and metabolically independent process that involves passive biological materials such as biomass [44]. The use of biomass with industrial microbial or agricultural wastes for synthesizing metal nanoparticles contained within wastewater represents an effective, innovative, and sustainable method of waste management.

Bioreduction is very important in both biosorption and bioaccumulation processes. It enhances ion metal biotransformation activity by reducing microorganisms; for example, Chromium (VI) can be reduced to nontoxic Chromium (III) using Shewanella alga, Ochrobactrum, Bacillus, and others [45]. Various researchers have assessed the use of microorganisms to transform Chromium (IV) to Chromium (III) [46–49].

Beside living and dead biomass, cell-free extracts are also used as bioreduction agents to produce metal [50]. Cell-free extract is used to give an enzyme and protein molecules capping to the metal salts to reduce them to metal nanoparticles. The cell-free extract from dead biomass produces a higher number of nanoparticles than living cells, dead biomass, and cell-free living cells [51]. The metabolically independent production process yields better morphology of nanoparticles than the metabolism dependent production [52]. This makes it possible to control the size and shape of nanoparticles by adjusting the amount of metal salt and the nonmetabolic bioreduction agents [53] and by controlling the pH of the aqueous solution [54]. The rate of bioreduction varies according to temperature; the higher the temperature, the higher the rate of bioreduction [55].

Even though the biosorption process is limited, especially in its ability to alter the metal valence state, there are some arguments on the selection of biomass, as reviewed by Nguyen et al. [16]. Beside the availability and cost effectiveness of biomass, selection also depends on the binding capacity of the biomass and selectivity for heavy metals. This will allow the development of a full-scale biosorption process.

Several characterization methods can be used to confirm the presence of metallic nanoparticles. These include transmission electron microscopy (TEM), Fourier transform infrared spectroscopy (FTIR), X-ray diffraction (XRD), energy dispersive X-ray spectroscopy (EDX), field emission scanning electron microscopy (FESEM), energy dispersive X-ray fluorescence spectrometry (EDXRF), and vibrating sample magnetometry (VSM) [56]. The crystallinity of nanoparticles can be investigated by numerous methods such as by XRD, for example, of iron oxide nanoparticles (Fe_3O_4 NPs) [56] and Silver nanoparticles (AgNPs) [57].

Important features that lead to the production of optimum nanoparticles are their optical properties, surface plasmon resonance (SPR) [58], and effective scattering phenomena [50]. SPR can be characterized by UV-Vis spectroscopy, while surface enhanced Raman scattering can be used to characterize scattering features. A number of groups have presented the ideal characteristics for Gold nanoparticles, where the UV-Vis spectrometer indicates the peak around 450 to 560 nm [50, 51, 59].

The biosynthesis process promises a green, safe, cost effective, and sustainable method of producing nanoparticles. The demands for heavy metal industrial activity have led to continuous bioremediation processes, which may lead to the depletion of both microorganisms and dead biomass. In order to reduce costs, it is important that we identify a method of recycling both dead and living microorganisms through the use of immobilization and desorption techniques [12]. The implementation of these techniques has enhanced bioremediation technology through the generation of recycled metal nanoparticles for further application.

The biological approaches used for bioremediation applications are always related to geomicrobiological process [4]. The metal-accumulating mechanisms involved in nanoparticle formation create metal-mineral-microbe interactions.

Metal Recovery from Heavy Metal Industry Effluent

Metal resources are nonrenewable; as such, the recovery of metal from industrial waste water could provide a means of maintaining the supply of heavy metals [10]. Scientists currently predict that a number of critical metal elements, such as Zinc, Silver, and Gold, will be depleted within the next 50 years if the current rate of consumption is maintained [60]. As a result of the rapid growth of the world economy, the consumption rate of these critical elements is increasing. As discussed in Section 2.1, several conventional remediation methods offer the potential to recover some of these metals from waste products, for example, chemical precipitation, membrane filtration, electrochemical treatment, biobleaching, and adsorption processes.

The use of a bioremediation process that employs a biosorption-desorption mechanism using natural biomass may represent one viable form of environmentally friendly bottom-up nanoparticles synthesis that could help to avoid a metal source depletion crisis [4, 61]. The definition of biosorption-desorption is discussed in Section 2.3. The quality of the recovered metals or nanoparticles depends on various aspects, such as types of biosorbents, type of metals, ion, pH, and temperature. The synthesis of high quality metal nanoparticles is a field of research that is generally discussed by groups outside bioremediation research. See Sections 2.3 and 2.4 for a comprehensive discussion of nanoparticle biosynthesis.

Various researchers have proposed the use of bioremediation to recover raw materials from waste [4, 61, 62]. In their review of heavy metal removal, Purkayastha et al. [63] compared Cd (II) recovery efficiency between conventional and contemporary methods. The contemporary method of biosorption is widely trusted and is the most popular and frequently used method of heavy metal removal and recovery. It also has the potential to be both a low-cost and environmentally friendly bioremediation process. From [63], a Cd(II) recovery of nearly 100% efficiency was demonstrated through the use of sulfide precipitation and biosorption processes. The conventional method of sulfide precipitation has the capability to recover Cd(II) up to 99.9% from its initial concentration [64]. Similar capability in Cd(II) recovery could also be obtained via a biosorption-leaching method, up to 98.89% from the initial concentration by using blackgram husk with 0.1 M HCl as a leaching agent [65]; the use of sawdusk obtained from mulberry wood with 1.5 M HCl has achieved Cd(II) recovery capability up to 92.79% [66], while Cd(II) recovery of up to 98.7% was obtained by using an Annona squamosa-based absorbent with 0.1 M HCl [67]. A Cd(II) recovery of up to 82% was obtained by using the biomass of Pseudomonas aeruginosa with 0.1 M HCl [68]. The recovery of Cu(II) of 100% was obtained by using volcanic rock matrix-immobilized Pseudomonas putida cells with surface-displayed cyanobacterial metallothioneins at pH 2.35 [69]. 100% recovery of Cu(II) was also obtained by using the activated sludge at pH 1 [70]. Over 90% recovery of Cu and Pb was obtained at pH $\leq$ 2 through the use of an indigenous isolateEnterobacter sp.J1 [71]. Liu et al. [72] compared the recovery efficiency between biosorption-leaching and biosorption-pyrolysis technology when recovering Pb from an aqueous solution. A Pb recovery of up to 94.9% was achieved in HCl solution, while 98.2% was achieved via the fast pyrolysis process.

Heavy metals can also be removed and recovered through bioleaching, which utilizes microorganisms as reduction agents [62]. The efficiency of the recovery process depends on the ability of the microorganisms to transform the solid compound within the contaminated soil into a soluble substance that can be extracted and recovered.

Several methods can be applied to achieve physical synthesis of metallic nanoparticles. These include attrition and pyrolysis. However, the process involves a significant conversion of energy to maintain the high pressure and temperature required during the synthesis process [73]. Top-down synthesis in the physical approach involves methods such as thermal decomposition, diffusion, irradiation, and arc discharge.

Chemical procedures are generally low-cost and can process high volumes; however, they typically involve the use of toxic solvents and generate hazardous byproducts. Examples of a bottom-up synthesis using a chemical approach include the seeded growth method, the polyol synthesis method, electrochemical synthesis, and chemical reduction. Scientists have successfully used a chemical reduction method to reduce a metal particle to nanoparticles using chemical agents such as Sodium borohydride or Sodium citrate.

Biological methods in the synthesis of metallic nanoparticles are becoming increasingly popular as they are low-cost, nontoxic, and environmentally benign. Synthesis using biological methods, especially those involving plants, can actively reduce metal ions in a biocompatible way where they can secrete functional biomolecules. The biological agents that are used in the biological approach to the synthesis of nanoparticles involve a variety of microbes [58, 74] including bacteria [73], fungi [75, 76], yeast, and plants [77].

Biosynthesis as an Approach to Producing Nanoparticles during a Bioremediation Process

Various review papers deal with the relationship between the bioremediation process and the production of metal nanoparticles [4, 61]. The main process that is used for both bioremediation and biosynthesis of nanoparticles is biosorption [8, 11]. Nanoparticle biosynthesis can be mediated by the biomass of plants and microorganisms (bacteria, algae, fungi, and yeast) through biosorption.

Biosynthesis is a bottom-up approach that uses biological molecular size entities to form nanoparticles [78]. Various biosynthetic methods are used to produce stable metal nanoparticles [79]. The biosynthesis of Gold, Silver, Gold-Silver, Platinum, Palladium, silica, alloy, Titanium, zirconia, Selenium, and Tellurium nanoparticles has already been reported [58]. Organisms, both unicellular and multicellular, have demonstrated unique potential in the environmentally friendly production and the accumulation of nanoparticles of different shapes and sizes that can be utilized for different commercial applications. Biosynthesis approaches for the production of metallic nanoparticles were reviewed by various groups [79]. Kulkarni and Muddapur [56] published a review on single-step biosynthesis mechanism of metal and metal oxide nanoparticles through the use of plants and microorganisms. The biomolecules present in plant extracts can be used to reduce metal or metal oxide ions to nanoparticles. This approach is low-cost, nontoxic, and environmentally benign and it allows the size of the nanoparticles to be controlled. The methods of nanoparticle characterization are used to understand the mechanism of particle formation and determine its future application [79]. In the biosynthesis of nanoparticles, different major factors, such as pH, temperature, concentration of metal ions, and concentration of extracts, influence the process of reducing metal ions to metal nanoparticles [57, 80].

Plant-mediated Biosynthesis of Nanoparticles

Plants have demonstrated a better ability to mediate nanoparticle synthesis than other methods and they offer a number of further advantages [78]. During the biosynthesis process, plants as biological agents act as reducing and capping agents [57]. Every part of the plant can be used for nanoparticles including leaves, flower, seeds, stems, fruits, latex, and calli. Furthermore, biomass from dead and dried plants can also be used for the synthesis of nanoparticles [81].

Research in plant-mediated biosynthesis of nanoparticles that has been conducted over the past 10 years (2003–2012) was reviewed extensively by Mittal et al. [74]. Over 50 different types of plant extracts have been used to synthesize metal nanoparticles and these have mostly been employed to produce Silver and Gold nanoparticles and alloys of different sizes and shapes [74]. A review of Iron, Zinc oxide, Selenium, Silver, and Gold nanoparticles that were mediated by plants, marine plants, and some microorganisms was published by Kulkarni and Muddappur [56]. Table 2 displays the results of a recent project which was not reviewed by Mittal et al. [74] and Kulkarni and Muddappur [56] in their review papers.

Table 2: Plant extract-based biosynthesis of metal or metal-based ion nanoparticles

Plant/plant part extract	Scientific name	Metal NPs	NPs size (nm)	Reference
Leaf	Lemon	Se	60–80	[82]
Leaf	Green tea	Fe ion		[83]
Leaf	Ecliptaprostrata	TiO	36–68	[84]
Leaf	Cinnamomum tamala	Au/TiO$_2$	8–20	[85]
Plant	Cacumen platycladi	Pt	2.4 ± 0.8	[86]
Leaf	Ocimum tenuiflorum (Tulasi)	Ag	7–15	[87]
Leaf	Chenopodium murale	Ag	30–50	[88]
Gum olibanum	Boswellia serrata	Ag	7.5 ± 3.8	[89]

Leaf	Cissus quadrangularis Linn	Ag	15–23	[90]
Leaf	Aloe	Ag	20	[91]
Fruit	Tribulus terrestris	Ag	16–28	[92]
Leaves	Stevia rebaudiana	Ag	2–50	[93]
Leaf	Artemisia nilagirica	Ag	70–90	[94]
Root	Morinda citrifolia	Ag	30–55	[95]
Leave	Rhizophora apiculata	Ag	19–42	[96]
Leaf	Prosopis juliflora	Ag	35–60	[97]
Leaf	Olive	Ag	20–25	[98]
Fruit	Terminalia chebula	Ag	25	[99]
Coir	Cocos nucifera	Ag	23 ± 2	[100]
Leaf	Malva parviflora	Ag	19–25	[101]
Leaf	Mangifera indica	Ag	20	[102]
Peel	Mangifera indica Linn (Mango)	Ag	7–77	[103]
Leaf	Origanum vulgare	Ag	136 ± 10.09	[104]
Leaf	Pepper	Ag	5–60	[105]
Leaf	Coccinia grandis	Ag	20–30	[106]
Leaf	Catharanthus roseus Linn G. Donn	Ag	27–32	[107]
Peel	Citrus unshiu	Ag	5–20	[108]
Plant	Scutellaria barbata D. Don	Au	5–30	[59]
Seed	Benincasa hispida	Au	10–30	[55]
Pod	Gymnocladus assamicus	Au	4.57 ± 0.23–22.57 ± 1.24	[109]
Leaf	Piper betle	Au	50 (mean size)	[110]
Pulp	Beta vulgaris	Au	Nanorod (25 nm) Nanowire (30 nm)	[111]
Leaf, stem, root	Ipomoea carnea	Au	3–100	[112]
Marine plant	Sargassum muticum	Au		[113]
Glucan of mushroom	Pleurotus florida	Au	5.33–18	[114]
Plant	Zingiber officinale	Au	5–15	[115]
	Crocus sativus	Au	11–20	[116]
Leaf	Hibiscus rosasinensis	Au		[117]
Leaf	Green tea	Au	20	[118]
Flower	Rosa damascena	Au, Ag	10–30	[119]

Of the reviewed projects (which were all conducted in the last 10 years before 2014), the majority of the plant extracts employed were derived from the leaf part of the plant. In order to create a sustainable method of recycling metal industrial effluent through the use of bioremediation technology, green and low-cost bioreduction agents from food industrial waste biomass should be used. Fruit peel, tea leaves, seeds, and flowers are among large-scale food industrial waste products and agricultural waste that should be reused to obtain a sustainable resource that can have a nanotechnology application.

The development of Gold and Silver nanoparticles is highly attractive to researchers due to the nobility of these metals and their wide range of applications, especially in biomedical and biochemistry fields. Remediating the waste produced by the Gold and Silver mining industry and its leachates through the use of food industrial waste not only will clean the water or soil of metal waste, but also could reduce the cost of the process.

Microorganism-mediated Biosynthesis of Nanoparticles

Bacteria-mediated Biosynthesis of Nanoparticles

The strong relationship between bioremediation technologies and the bacteria-mediated biosynthesis of nanoparticles through the application of biosorption was discussed by various researchers [120, 121]. Kulkarni and Muddappur [56] reviewed recent research on the biosynthesis of metal nanoparticles through the use of bacteria as reduction agents. Table 3 lists the types of bacteria that synthesize metal nanoparticles. Most such syntheses are extracellular binding metal nanoparticles to bacterial biomass. The bacterial biomass, such as Pseudomonas aeruginosa and Aeromonas hydrophila, is obtained at low cost from waste from bacterial synthesis in the plastic industry, polyhydroxyalkanoates [122]. Due to the biodegradability of these materials, polyhydroxyalkanoates are desirable and will offer continuous production on a large scale [123]. Thus, the production of both P. aeruginosa and A. hydrophila biomass will continuously be available and offer access to low-cost Gold and Zinc oxide nanoparticle biosynthesis, respectively, especially with intention to recover these metals from the industrial effluent produced by the metals industry. The Selenium and Titanium oxide nanoparticle biosynthesis could also be continuously obtained, since Bacillus subtilis biomass is available on a continuous basis. Bacillus subtilis-based enzymes are in high demand in a number of consumer chemical production industries such as those that produce cleaning products, paper and textiles, food, and pesticides [124].

Table 3: Bacteria-mediated biosynthesis of nanoparticles

Metal/metal oxide	Microorganisms		Nanoparticle size (nm)
Au	Geobacillus sp. [125]		5–50
	Klebsiella pneumoneae [126]		
	Escherichia coli [127, 128]	Extracellular	17–32, 5–70 (uniform at 2.2)
	Magnetospirillum gryphiswaldense MSR-1 [54]	Extracellular	10–40
	Pseudomonas aeruginosa [129]	Extracellular	15–30
	Rhodopseudomonas capsulate [130]	Extracellular	10–20
	Micrococcus luteus [131]	Extracellular	6 nm and 50 nm
	Stenotrophomonas [132]	Extracellular marine source	10–50
Ag	E.coli [133]		
	Lactobacillus sp. [134]		10–25,
	Bacillus licheniformis [135]		2–100
	Streptomyces hygroscopicus (BDUS 49) [136]	Live cells from sewage	20–30
	Corynebacterium glutamicum (0) [120]		5–50
	Streptomyces sp. BDUKAS10 [137]	Extracellular	21–48
	Bacillus cereus [138]		4-5
	Bacillus amyloliquefaciens LSSE-62 [139]	Intracellular	14.6
	Stenotrophomonas [132]		40–60
Se	Klebsiella pneumoneae [9, 133]		100–550
	Zooglea ramigera [140]	Extracellular	30–150
	Bacillus subtilis [141]	Extracellular	50–400
Ag_2O	Lactobacillus mindensis [142]		2–20

Ti	Lactobacillus sp. [134]		10–70
TiO$_2$	Bacillus subtilis [143]		66–77, 10–30
Cu$_2$S	Streptomyces sp. [144]		100–150
	Desulfovibrio desulfuricans [121]		20–30
Zinc nitrate	Streptomyces sp. [144]		100–150
ZnO	Calotropis gigantean [145]		
	Lactobacillus sporogenes [146]		5–15
	Lactobacillus plantarum VITES07 [147]		7–19
	Aeromonas hydrophila [148]		57.72
CdS	Rhodopseudomonas [149]	Intracellular	8.01 ± 0.25

Many other forms of bacteria biomass could be obtained from industrial waste and subsequently be used for the low-cost, environmentally friendly, and sustainable bioremediation of metal industrial effluents.

Fungi-mediated Biosynthesis of Nanoparticles

The fungi-mediated biosynthesis of Silver nanoparticles was reported by Duran et al. [79]. Uniformity in terms of size and shape of the Gold nanoparticles formed through the use of Aspergillus oryzae var. viridis (waste industrial fungal biomass from industry) ranged between 10 and 60 nm [51]. Mishra et al. [52] reported the potential of an industrially important fungus, Penicillium rugulosum, for the synthesis of Gold nanoparticles. Reduction of Silver nitrate to Silver metallic nanoparticles is currently an established routine in laboratories worldwide. Vigneshwaran et al. [150] investigated the use of white rot fungus Phaenerochaete chrysosporium for the extracellular synthesis of Silver nanoparticles. The mycelium of P. chrysosporium was found to reduce Silver nitrate to metallic silver nanoparticles. Utilization ofCoriolus versicolor to reduce Silver nitrate to Silver metallic nanoparticles was reported by Sanghi and Verma [151]. The formation of Silver nanoparticles through an extracellular cell wall reduction process of Silver nitrate by using Aspergillus flavus was successfully performed by Vigneshwaran et al. [152]. Silver nanoparticles with a size of 8.92 ± 1.61 nm were obtained in this study. Silver metallic nanoparticles that were 5–25 nm in size were obtained through extracellular biosynthesis by using Aspergillus fumigatus [153]. Table 4shows the fungi-mediated biosynthesis of metals nanoparticles for noble metals.

Table 4: Fungi-mediated biosynthesis of metal nanoparticles

NPs	Name	Binding location	Size of NPs (nm)
Ag	Phaenerochaete chrysosporium [150]	Extracellular	50–200
Ag	Aspergillus flavus [152]	Extracellular	8.92 ± 1.61
Ag	Coriolus versicolor [151]	Intra-/ extracellular	
Ag	Penicillium brevicompactum WA2315 [154]	Extracellular	23–105
Ag	Cladosporium cladosporioides [155]	Extracellular	10–100
Ag	Candida albicans [50]		5–30
Ag	Neurospora crassa [156]	Intra-/ extracellular	11
Ag	Fusarium oxysporum [157]	Extracellular	5–15
Pt	Fusarium oxysporum [158]	Extracellular	5–30
TiO$_2$	Aspergillus flavus [159]	Extracellular	62–74
TiO$_2$	Humicola sp.[160]		
ZnO	Candida albicans [161]		
Au	Aspergillus oryzae var. viridis [51]		30–400
Au	Sclerotium rolfsii [53]		25
Au	Penicillium rugulosum[52]		20–80
Au	Cylindrocladinum [162]		5–35
Au	Candida albicans [50]		5–30
Au	Neurospora crassa [156]	Intra-/ extracellular	32

Algae-mediated Biosynthesis of Nanoparticles

Abdel-Aziz et al. [88] successfully obtained Gold nanoparticles that were 3.85–77.13 nm in size through the use of an extract of the Galaxaura elongata algae. The use of G. elongata to obtain the nanosize Gold nanoparticles demonstrated the high effectiveness of this method compared to the use of E. coli and K. pneumonia, which yielded Gold nanoparticles of size 13.5 and 13 nm, respectively. Arockiya et al. [163] obtained Gold nanoparticles of a size ranging between 18.7 and 93.7 nm through the use of Stoechospermum marginatum. The obtained nanoparticles demonstrated potential to be used as antibacterial agents.

Yeast-mediated Biosynthesis of Nanoparticles

Au nanoparticles have been obtained through the use of the Saccharomyces cerevisiae AP22 yeast and CCFY-100 through accumulation processes inside the cells [164], while Ag nanoparticles with an average size of 19 ± 9 nm were obtained through the use of the Saccharomyces cerevisiae BU-MBT-CY1 yeast [165].

Hyperaccumulation of Metals for Phytomining Applications

Accumulation of metal by plant species for soil remediation was first used as early as 1983, and it advanced to an accepted metal mining technology in recent years [34]. Metal accumulating species can be used for phytoremediation (removal of contaminants from soils) (as discussed in Section 2.4) or for phytomining, which involves growing plants to harvest metals (this will be discussed in this section). In addition, many of the metals that can be hyperaccumulated also provide essential nutrients; food fortification and phytoremediation might be considered two sides of the same coin [166]. Many heavy metals are essential or beneficial as micronutrients for the growth and metabolism of microorganisms, plants, and animals but are dangerous if found in high concentrations. These include Cobalt (Co), Copper (Cu), Iron (Fe), Manganese (Mn), Molybdenum (Mo), Nickel (Ni), and Zinc (Zn). On the other hand, some heavy metals do not seem to be essential because there is no demonstrated biological or physiological function (yet). These include Lead (Pb), Arsenic (As), Cadmium (Cd), and Mercury (Hg). Several reasons and hypotheses exist concerning why some plants hyperaccumulate metals. The first is that the metals [166] provide the plants with a physiological strategy that protects them from herbivore attack through feeding deterrence and from pathogen attack through their toxicity [34, 166]. Sagner and coworkers [167] observed the repellent effect that plant sap had on the fruit fly Drosophila melanogaster, indicating that, in hyperaccumulating plants, Nickel serves as an agent to prevent predation. Furthermore, the defensive enhancement increases with increasing metal concentration [168]. Hyperaccumulator species are distributed across a wide range of distantly related families, showing that the hyperaccumulation trait has evolved independently more than once under the spur of selective ecological factors. The ability of plant species to accumulate heavy metals is of interest in multidisciplinary fields as it holds potential for the development of technologies that are of human and environmental concern.

Hyperaccumulation of Metal by Plants

Plants have the ability to accumulate metals from the soil in their tissue during growth and development. The hyperaccumulation process offers a new, environmentally friendly method of producing metal nanoparticles [34]. The term "hyperaccumulator" describes plants that have the ability to grow on metalliferous soils and to accumulate extraordinarily high amounts of heavy metals that are far in excess of the levels found in the majority of species in their aerial organs, without suffering phytotoxic effects [169]. Chaney [170] proposed the idea of using plants that hyperaccumulate metals to selectively remove and recycle excessive metals in soil. Hyperaccumulator plants have been defined as those that accumulate metal in concentrations that are 10–100 times more than those found in normal plants [170]. According to Reeves [171], there are 440 hyperaccumulator plant species, of which 75% are Nickel hyperaccumulators. Other plant species accumulate metals like Cadmium, Arsenic, Manganese, Sodium, Thallium, and Zinc. When metal ions have been taken up and are concentrated in the tissue of the hyperaccumulator plants, the biomass may be harvested, dried, and burnt to ash for recycling as a bio-ore [172, 173] or stored for later use.

Wei and coworkers [174] summarized the main characteristics of hyperaccumulator plants as follows:

- accumulation property, that is, the minimum concentration in the shoots of a hyperaccumulator; for As, Pb, Cu, Ni, and Co, it should be greater than $1000\,mg\,kg^{-1}$ dry mass, for Zn and Mn $10,000\,mg\,kg^{-1}$, for Au $1\,mg\,kg^{-1}$ and Cd $100\,mg\,kg^{-1}$, respectively [175];

- translocation property; that is, elemental concentrations in the shoots of a plant should be higher than those in the roots [176, 177];

- enrichment property; that is, enrichment factor EF (concentration ratio of plant to media) in shoots of plants should be higher than those in roots [178];

- tolerance property; that is, a hyperaccumulator should have high tolerance to heavy metals [178]. The tolerance mechanism in plants that hyperaccumulate Cd was explored from root morphology by Wei and coworkers [174].

Hyperaccumulation by plants includes the uptake of heavy metal through "root-to-shoot." The majority of the heavy metal is absorbed from the soil by the root and then detoxified by chelation in the cytoplasm or stored in vacuoles. Hyperaccumulators rapidly and efficiently translocate the element to the shoot via the xylem. Rascio and Navari-Izzo [166] described three basic characteristics by which hyperaccumulator plants can be distinguished from nonaccumulation plant species:

- a much greater capability to extract heavy metals from the soils,
- a faster and effective root-to-shoot translocation of metals, and
- a much greater ability to detoxify and sequester huge amounts of heavy metals in the leaves.

Plants Species in Hyperaccumulation

Many species of plant can hyperaccumulate heavy metals. These includeAsteraceae, Brassicaceae, Caryophyllaceae, Cyperaceae, Cunouniaceae, Fabaceae, Flacourtiaceae, Lamiaceae, Poaceae,

Violaceae, Sapotaceae, and Euphobiaceae. The Brassicaceae family has the largest number of taxa, consisting of about 11 genera and 87 species. Members of the family Bracecaceaeaare well known as hyperaccumulators of Nickel (genera Thalaspi and Alyssum), Cadmium, and Zinc (Thalaspi caerulescens, T. praecox, T. geosingese, and Arabidopsis halleri). According to research performed by Sagner et al. [167] and Jaffre et al. [179], the highest concentration of Nickel in latex has been recorded in different parts (latex, leaves, trunk bark, twig bark, fruits, and wood) of the Sebertia acuminata (Sapotaceae) tree from New Caledonia. More taxa of plants species are accumulators of Nickel (more than 75%) than of any other metal, and only five species accumulate Cadmium [166]. Several heavy metal hyperaccumulating species that belong to the Thlaspigenus have been studied, among them Thlaspi caerulescens hyperaccumulating Zn/Cd [180], Thlaspi rotundifolium ssp. cepaeifolium hyperaccumulating Pb [173], and T. praecox Wulf hyperaccumulating Zn [181]. The hyperaccumulator plant Alyssum murale can accumulate Ni up to $20,000\,mg\,kg^{-1}$ dry weight, and a biomass of 10,000 kg/ha can be harvested per year [182]. Serpentine soils contain Ni at concentrations between 1000 and $7000\,mg\,kg^{-1}$. These concentrations are well below the exploitation threshold required by traditional mining ($30,000\,mg\,kg^{-1}$) but are sufficient to allow hyperaccumulating plants to extract and accumulate Ni [175]. A selection of hyperaccumulator plant species and their threshold concentrations is given in Table 5.

Table 5: Hyperaccumulators in phytomining. Source: van der Ent et al. [183] with modifications, together with the metal concentration in selected hyper-accumulators

Metal	Number of hyperaccumulator species recorded	Hyperaccumulation threshold ($mg\,kg^{-1}$)	Selected hyperaccumulator species	Concentration of metals (mg/ kg d.w.)
			Sebertia acuminate[167, 179]	
			Streptanthus polygaloides [184, 185]	
			Alyssum bertolonii[186, 187]	13400

Nickel	450	1000	Berkheya coddii [187]	17000
			Thlaspi geosingense[188]	
			Alyssum tenium [19]	
			Alyssum troodii [19]	
Cobalt	30	300	Haumaniastrum robertii [189]	10200
Copper	32	300	Helianthus annuus L. [190]	
			Haumaniastrum katangense [169]	8356
Zinc	12	3000	Thlaspi calaminare[181]	10000
			Thlaspi caerulescens	
Zinc and Cadmium			Sedum alfredii	
			Polycarpaea longiflora[19]	
			Allium sativum L. [191]	
Cadmium	2	100	Thlaspi caerulescens[19]	3000
			Solanum nigrum [192]	
			Rorippa globosa [174]	
Gold (induced hyperaccumulation)		1	Brassica juncea Barkley coddii [193,194]	10
Manganese	12	10000	Macadamia neurophylla [195]	55000
Lead	14	1000	Thlaspi rotundifoliumsubsp. [181]	8200
Thallium	2	100	Biscutella laevigata Iberis intermedia [193,196]	4055

Lin and coworkers [190] examined the different concentrations of Copper sulphate in the growth and accumulation of Cu^{2+} in different parts, such as in the root, hypocotyl, cotyledon, and leaf, of the sunflower (Helianthus annuus L.) plant. They found that the concentration of Cu^{2+} was higher in the roots than it was in other parts of the Helianthus annuus and concluded that these plants have the potential ability to accumulate Copper. Copper is a catalytic cofactor of enzymes and is necessary in the normal growth and development of many plants species. Juárez-Santillán and coworkers [197] carried out an experiment that aimed to identify Manganese accumulation in the plants growing in the mining zone of Hidalgo, Mexico. They used eight species of accumulator plants and found a Mn content in the substrate ranging from 11,637 to 106,104 mg kg^{-1}dry weight, with the concentrations being between 2 and 21 times higher than the phototoxic level of 5,000 mg kg^{-1} dry weight according to Alloway [198]. Jiang and coworkers [191] investigated the effects of Cadmium chloride concentration, uptake, and accumulation of Cd^{2+} using the hyperaccumulator garlic Allium sativumL. Their results showed that the concentration of Cadmium was higher in the roots than it was in the shoot and bulb.

Types of Hyperaccumulation

There are two types of hyperaccumulation, natural (using plants) and induced (adding chemicals). Natural hyperaccumulation occurs when plant species use their physiological ability to accumulate the state of heavy metals as a normal function of their growth, while induced hyperaccumulation is performed by adding chemicals to the soil in order to manipulate the soil-plant environment. Several pieces of research have covered hyperaccumulation [34]. The majority of the discussion in the existing reports concentrates on natural hyperaccumulation. Anderson and coworkers [193] attempted to induce plants to hyperaccumulate Au by adding ammonium thiocyanate to substrate, and their results revealed that Indian mustard (Brassica juncea) accumulated up to 57 mg kg^{-1}. In this research, the hyperaccumulation of Gold was defined as accumulation greater than 1 mg kg^{-1} based on normal concentration in plants of 0.01 mg kg^{-1} [199,200].

Hyperaccumulation of Metal by Others

Aquatic macrophytes, such as Eleocharis acicularis, hold great potential for phytoremediation of water contaminated with multiple heavy metals from mine tailings, mine drainage, and water [201, 202]. Ha and coworkers [203] completed a laboratory investigation in which they studied the ability of dwarf hair grass (Eleocharis acicularis) to accumulate Indium, Silver, Copper, Cadmium, Lead, and Zinc and assessed phytoremediation and phytomining. Their results indicated that Eleocharis acicularis does have the ability to accumulate metals; the concentrations of metals in the roots of the plant were as follows: Indium 477 mg kg^{-1}, Silver 326 mg kg^{-1}, Copper 575 mg kg^{-1}, Cadmium 195 mg kg^{-1}, Lead 1120 mg kg^{-1}, and Zinc 213 mg kg^{-1} dry weight after 15 days exposure, and they concluded that there was the potential to extract Indium and Silver from these plants and that they can be used for phytomining.

How Organisms Accumulate Heavy Metals

As discussed above, bioaccumulation involves the accumulation of metal from the soil or water and is always performed by living microorganisms such as bacteria, fungi, algae, and yeast [8, 10]. The microorganisms take up the metal into the cell across the cell membrane [11]. Understanding the mechanism of biosorption and the structure of the cell wall or cell membrane is important to identify an appropriate method of removing metals from polluted water or soil. Microbes, including eukaryotes and prokaryotes, interact with metals and minerals in natural and symbiotic association with each other and other organisms. They can alter the physical and chemical condition and have an impact on the growth, activity, and survival of microbials [4]. Chmielewská and Medved [204] confirmed the high Ni^{2+} bioaccumulation ability of Cladophora glomerata. Furthermore,Microspora also might have potential to be useful in bioremediation. The mechanisms by which metal is bound to microbial biomass are divided into three types:

- intracellular accumulation, where the process requires live cells,
- sorption or complex formation on the cell surface for both living and dead cells, and

- extracellular accumulation or precipitation where the process may require viable cells [205, 206].

Bioaccumulation via growing cells is a potential technique for removing heavy metals from a food environment. The heavy metals can be both biosorbed onto the cell surface and passed into the cell across the cell membrane through the cell metabolic cycle [207].

The uptake of metal ions by living and dead cells consists of two methods; either the metal ions bind to the surface of the cell wall and extracellular material or the metal is absorbed into the cell across the cell membrane, which is referred to as intracellular uptake, active uptake, or bioaccumulation [5]. The first method occurs in both living and dead cells in biomass, while the second method, which is dependent on the plants' metabolism, occurs only in living cells. As such, both living and dead cells are capable of metal adsorption [208]. The use of dead biomass is preferred to living matter due to the absence of toxicity limitations, growth media and nutrient requirements, and the high capacity of the binding metals. Metal uptake is also facilitated by the production of metal-binding proteins in living cells.

A less energy dependent approach for bioaccumulating Gold nanoparticles was successfully performed through the use of two different strains of yeast, Saccharomyces cerevisiae AP22 and CCFY-100 [164]. The ion Gold (III) is reduced to Gold (0) and the nanoparticles accumulate inside the cell nucleolus. Narayanan and Sakthivel [162] successfully synthesized nano-Gold composite by using the fungus Cylindrocladium floridanum via an intracellular accumulation process. This study introduced the use of microbially matrixed Gold nanoparticles as heterogeneous catalysts to control the accumulation of the nano-Gold composite.

Phytomining of Various Metals

Phytomining has emerged as an environmentally friendly technology that uses plants to extract heavy metals from a given substance [173, 193]. This technology involves growing and harvesting metal-accumulating plants on appropriate sites and treating the biomass to recover the metal. Phytomining describes the bioharvesting of metals from high biomass crops grown in soil substrates, particularly those associated with subeconomic mineralization. Phytomining involves growing high-biomass plants that accumulate high metal concentrations. Phytomining technology is perhaps the most feasible, lowcost, and environmentally friendly alternative to conventional mining methods as it allows the economic exploitation of mineralized soils that are thought too metal-poor for direct mining operations. The technique of phytomining involves growing a crop of a metal-hyperaccumulating plant species, harvesting the biomass, and burning it to produce a bio-ore [193]. Pythomining has been suggested as a potential alternative to recovering metal for other applications. In this situation, several studies have attempted to recover metals, such as Nickel [189, 209, 210], from mineralized soils or waste rock. Phytomining employs hyperaccumulating plants to extract valuable metals from the substrate. Many metals, such as Nickel, Cadmium, and Manganese, occur naturally in hyperaccumulator plant species because most metals are bioavailable in the soil solution in which the plants grow [175].

Phytomining offers the possibility to exploit metals derived from low-grade ore, overburdens, mill tailings, or mineralized soil, which are uneconomic to extract by conventional methods [19]. The metal content of a bio-ore is usually much greater than that of a conventional ore and requires less storage space, despite the lower density of a bio-ore [173, 193]. Besides that, phytomining can also be used for remedial purposes to extract high concentrations of toxic metals, such as Nickel and Thallium, from mine tailings and other metal contaminated areas. Phytomining not only produces ingots of Gold but, interestingly and importantly, this technique can also be used to produce Gold nanoparticles, which is of huge potential to nanotechnology industries [211–213]. Additionally, phytomining provides a source of income from the sales of Carbon dioxide credits. Phytomining restores mined degraded land through the planting of hyperaccumulator species more quickly than natural revegetation which may take decades or even hundreds of years because it is dependent on animals and windborne seedlings. Phytomining also provides a potential energy storage that can be utilized to generate thermal energy, which is a cheap and renewable resource.

Phytomining of Nickel

Numerous studies have examined the use of Ni phytomining. The first field trials involving phytomining were carried out at the US Bureau of the Mines, Reno, Nevada, using the Ni hyperaccumulator Streptanthus polygaloides [184]. The findings of this initial research indicated that a yield of 100 kg/ha of sulfur-free Ni could be produced via phytomining [186, 193, 214]. Anderson and coworkers [193] tested the phytomining potential of Ni hyperaccumulator Alyssum bertolonii from Italy and Berkheya coddiifrom South Africa. In situ, an experiment for second field trials of Nickel phytomining using the Ni hyperaccumulator Alyssum bertolonii was performed in Tuscany, Italy [186]. In this experiment, the plants were fertilized with N + P + K combinations over a period of two years. The results indicated that the biomass of dry matter increased threefold to 9.0 t/ha and this was gained with N + P + K without dilution of unfertilized Ni concentrations. The existing research does not show a correlation between the age of a plant and Ni content. The hyperaccumulator Alyssum bertolonii has the potential to be used in Nickel phytomining of ultramafic soils. The hyperaccumulating plant Alyssum murale can accumulate up to 20,000 mg kg^{-1} dry weight, and a biomass of 10,000 kg/ha can be harvested per year [182].

Phytomining of Gold

Gold is a valuable metal that is in high demand in commercial industries. It has a potential for phytomining where the aim of the operation is to yield an economic profit and/or exploit low-grade ore or mineralized soils that are too poor for conventional mining methods [215]. Conventional technology is generally unable to economically recover residual Gold, and, as such, phytomining offers a viable alternative method of recovering these valuable Gold resources [216]. Plant species do not naturally accumulate Gold; it needs to be made soluble in soils for enhancing uptake to occur. For this purpose, hyperaccumulation is induced by adding Sodium cyanide, thiocyanate, and thiosulphates. The residual Gold can be extracted using induced hyperaccumulation if the substrate were amenable to plant growth. The Gold concentration that can be induced into a plant is dependent on the Gold concentration in the soil in which the plant is growing. Induced hyperaccumulation to uptake Gold has been reported in numerous pieces of research. Anderson and coworkers [187] showed that approximately 2 mg kg^{-1} of Gold are needed to achieve 100 mg kg^{-1} of plant dry weight in Indian mustard (Brassica juncea) with adding ammonium thiocyanate (NH_4SCN) at different rates of 0, 80, 160, 320, and 640 mg kg^{-1} dry substrate weight in pots containing an artificial 5 mg kg^{-1} finely disseminated Gold rich material, analogous to natural, oxidized, and nonsulphidic ores. Msuya and coworkers [194] induced hyperaccumulation in the five root crops, carrot (Daucus carota), red beet (Beta vulgaris), onion (Allium cepa), and two cultivars of radish (Raphanus sativus) with NH_4SCN and ammonium thiosulphate [$(NH_4)_2S_2O_3$] in substrates containing 3.8 mg kg^{-1} Gold. The results showed that the roots in all five crops contained higher metal concentrations than their shoots.

Phytomining of Thallium

Thallium has the biggest economic potential after Gold, Platinum, and Palladium because it is relatively naturally unavailable. Leblanc and coworkers [196] discovered unusually high hyperaccumulation of Tl by Iberis intermedia Guersent and Biscutella laevigata L. growing over Lead/Zinc mine tailings in southern France. Tailings typically contained 15,000 mg kg^{-1} Zn, 5,000 mg kg^{-1} Pb, and, locally, up to 40 mg kg^{-1} Tl. Their results indicated that the content of Thallium in Iberis intermedia was more than 4,000 mg kg^{-1} in the whole plant dry weight with a biomass of 10,000 kg/ha, while in the Biscutella laevigata the content of Thallium was over 14,000 mg kg^{-1} with a biomass of 4,000 kg/ha. Biscutella laevigata at 4 t/ha has less than half of the biomass of Iberis intermedia but three times the Thallium mean concentration (mean 10 mg kg^{-1}). These results are similar to those reported in field trials in New Zealand, where a biomass of 10,000 kg/ha was found in Iberis intermedia by Anderson and coworkers [193]. The hyperaccumulator plant produces about 700 kg/ha of bio-ore and 8 kg of Thallium, which was worth US$ 2400 at a world price of US$ 300/kg. A crop biomass of 10 t/ha of Iberis intermedia was found from field observations in France and field trials in New Zealand and this indicates that there is a potential to phytomine Thallium if sufficiently large areas of contaminated soils are available and large-scale production advantages can be acquired [193].

Biomineralization

The collective process by which organisms form minerals is called biomineralization [217]. Metal-mineral-microbe interactions are basically microbial biomineralization processes [4]. More than 70 different minerals are produced by organisms via biomineralization [218]. The preparation of biosorption is also important for estimating the operational costs associated with phytomining.

THE POTENTIAL OF PRODUCTS DERIVED VIA BIOREMEDIATION FROM HEAVY METAL INDUSTRY EFFLUENTS FOR HIGH TECHNOLOGY APPLICATIONS

Nanotechnology Applications

Metal nanoparticles, such as Gold, Silver, and Platinum, have become fundamental building blocks of nanotechnology [219]. The properties, morphology, surface area, structure, and shape of nanoparticles make them ideal for preparing nanostructured materials and devices [75, 220, 221]. Metal nanoparticles are crystalline in nature [56, 57]. To control the size, shape, and stability of nanoparticles, different synthesis conditions are applied [56].

Nanotechnology is an interdisciplinary area of science, economy, engineering, and industry, and increasing attention is being placed on the production of nanoparticles. Nanotechnologies use different metal nanoparticles in a wide range of sizes and shapes. Nanoparticles (nanocrystals, Carbon nanotubes, etc.) are microscopic particles. They are usually between 1 and 100 nm in size in each spatial dimension and are considered to be the building blocks of the next generation of optoelectronics, electronics, and various chemical and biochemical sensors. A variety of nanomaterials are at various stages of research and development, each one possessing unique functionalities that are potentially applicable to the remediation of industrial effluents, groundwater, surface water, and drinking water. Silver nanomaterials are the most interactive in the application of nanotechnology. Silver nanoparticles can be used in the development of antibacterial water filters for the treatment of waters [222]. Tran and coworkers [223] reviewed Silver nanoparticles in depth and assessed synthesis, antibacterial effects, toxicology to humans and the environment, current applications, and future prospects. Silver nanoparticles are of great interest due to their optical, electrical, and antimicrobial properties [224].

Sustainable Green Metal Mining

In recent decades, techniques for mining metals using plants—phytomining—have offered scientists the possibility to extract metals from low-grade ores, overburdens, mill tailings, or mineralized soil that would be uneconomical to mine using conventional methods [19]. Plant hyperaccumulators accumulate large amounts of metals in their roots, such as Nickel and Thallium, and hence provide a potential route for soil remediation and for the recovery and reuse of the metal [193]. Phytomining not only produces ingots of Gold but, interestingly and importantly, this technique can also be used to produce Gold nanoparticles, which are of huge potential to nanotechnology industries [211–213]. The biological systems that are used in the synthesis of nanoparticles exhibit water-soluble and biocompatible properties that are essential for many field applications, from medicine to electronics [76]. A variety of microorganisms, plants, and plant parts, including roots, stems, leaves, and bark, provide low-cost, energy efficient, and nontoxic methods of synthesizing nanoparticles. Heavy metals can be removed from contaminated soil and water via hyperaccumulation, either through phytoremediation or through phytomining, for later use as material productions in nanotechnology.

CONCLUSIONS

This work reviewed existing research that examines the potential of the bioremediation process for recovery of heavy metals from contaminated water and soil sources. The use of cheap and easily obtained biomass using this approach offers a low-cost and environmentally friendly solution and has the potential to be used for large-scale bioremediation. Biosorption-based bioremediation represents a low-cost and environmentally friendly method of metal recovery that may offer a means of avoiding metal depletion. The bioremediation process proposed could produce metal nanoparticles through biosorption-based biosynthesis, which also uses the biomass from plants and microorganisms. Metal nanoparticles are very important for high technology applications, such as biomedical applications, sensors, MEMS (microelectromechanical systems) and NEMS (nanoelectromechanical systems), catalysts, and antimicrobial agents. Phytomining is an environmentally friendly method of removing heavy metals from wastewaters and soil. This method is important, especially in areas in which the soil contains high levels of heavy metals.

Based on the past and present works of research in bioremediation and phytomining processes, the future of such attempts looks bright. Regarding the conversion of heavy metal effluents from waste material to a valuable resource for research, development, and industry, there is still a plenitude of basic research that needs to be performed to establish the basics, especially concerning which plants can be used on which soils with which pollutant concentration and in which environment (e.g., other plants and microbiome communities, latitude, and longitude).

The continuous implementation of both bioremediation and phytomining processes offers great benefits of green methods in various fields, such as environment, resources, economy, and human life. Comprehensive research and development for improvement of both processes promise a bright future for successful implementation of biomimetic approaches in resource management.

ACKNOWLEDGMENTS

This work was supported by the Fundamental Research Grant Scheme from the Ministry of Higher Education, Malaysia (project no. FRGS/1/2013/TK02/UKM/01/1).

REFERENCES

1. I. C. Gebeshuber, B. Y. Majlis, and H. Stachelberger, "Tribology in biology: biomimetic studies across dimensions and across fields," International Journal of Mechanical and Materials Engineering, vol. 4, no. 3, pp. 321–327, 2009.

2. W. H. Rulkens, R. Tichy, and J. T. C. Grotenhuis, "Remediation of polluted soil and sediment: perspectives and failures," Water Science and Technology, vol. 37, no. 8, pp. 27–35, 1998. · ·

3. C. Garbisu and I. Alkorta, "Review: basic concepts on heavy metal soil bioremediation," The European Journal of Mineral Processing and Environmental Protection, vol. 3, pp. 58–66, 2003.

4. G. M. Gadd, "Metals, minerals and microbes: geomicrobiology and bioremediation," Microbiology, vol. 156, no. 3, pp. 609–643, 2010. · ·

5. S. Bhatnagar and R. Kumari, "Bioremediation: a sustainable tool for environmental management—a review," Annual Review & Research in Biology, vol. 3, no. 4, pp. 974–993, 2013.

6. F. Fu and Q. Wang, "Removal of heavy metal ions from wastewaters: a review," Journal of Environmental Management, vol. 92, no. 3, pp. 407–418, 2011. · ·

7. M. A. Hashim, S. Mukhopadhyay, J. N. Sahu, and B. Sengupta, "Remediation technologies for heavy metal contaminated groundwater," Journal of Environmental Management, vol. 92, no. 10, pp. 2355–2388, 2011. · ·

8. A. Malik, "Metal bioremediation through growing cells," Environment International, vol. 30, no. 2, pp. 261–278, 2004. · ·

9. Churngold Group, Heavy Metal Contamination, 2014.

10. B. W. Atkinson, F. Bux, and H. C. Kasan, "Considerations for application of biosorption technology to remediate metal-contaminated industrial effluents," Water SA, vol. 24, no. 2, pp. 129–135, 1998.

11. R. Dhankhar and R. B. Guriyan, "Bacterial biosorbents for detoxification of heavy metals from aqueous solution: a review," International Journal of Advances in Science and Technology, vol. 2, pp. 103–128, 2011.

12. K. Vijayaraghavan and Y.-S. Yun, "Bacterial biosorbents and biosorption," Biotechnology Advances, vol. 26, no. 3, pp. 266–291, 2008. · ·

13. J. He and J. P. Chen, "A comprehensive review on biosorption of heavy metals by algal biomass: materials, performances, chemistry, and modeling simulation tools," Bioresource Technology, vol. 160, pp. 67–78, 2014. · ·

14. A. V. Bankar, S. S. Zinjarde, B. P. Kapadnis, T. Satyanarayana, B. N. Johri, and A. Prakash, "Management of heavy metal pollution by using yeast biomass," in Microorganisms in Environmental Management, pp. 335–363, Springer, Berlin, Germany, 2012. ·

15. T. Viraraghavan and A. Srinivasan, "Fungal biosorption and biosorbents," in Microbial Biosorption of Metals, P. Kotrba, M. Mackova, and T. Macek, Eds., pp. 143–158, Springer, Dordrecht, Netherlands, 2011.

16. T. A. H. Nguyen, H. H. Ngo, W. S. Guo et al., "Applicability of agricultural waste and by-products for adsorptive removal of heavy metals from wastewater," Bioresource Technology, vol. 148, pp. 574–585, 2013. · ·

17. S. M. Shaheen, F. I. Eissa, K. M. Ghanem, H. M. Gamal El-Din, and F. S. Al Anany, "Heavy metals removal from aqueous solutions and wastewaters by using various byproducts," Journal of Environmental Management, vol. 128, pp. 514–521, 2013. · ·

18. A. Bhatnagar and M. Sillanpää, "Utilization of agro-industrial and municipal waste materials as potential adsorbents for water treatment—a review," Chemical Engineering Journal, vol. 157, no. 2-3, pp. 277–296, 2010. · ·

19. T. V. Nedelkoska and P. M. Doran, "Hyperaccumulation of cadmium by hairy roots ofThlaspicaerulescens," Biotechnology and Bioengineering, vol. 67, pp. 607–615, 2000.

20. S. Babel and T. A. Kurniawan, "Low-cost adsorbents for heavy metals uptake from contaminated water: a review," Journal of Hazardous Materials, vol. 97, no. 1–3, pp. 219–243, 2003. · ·

21. P. N. Dave and L. V. Chopda, "Application of iron oxide nanomaterials for the removal of heavy metals," Journal of Nanotechnology, vol. 2014, Article ID 398569, 14 pages, 2014. ·

22. L. S. Acosta-Torres, L. M. Lpez-Marín, R. E. Núñez-Anita, G. Hernández-Padrón, and V. M. Castaño, "Biocompatible metal-oxide nanoparticles: nanotechnology improvement of conventional prosthetic acrylic resins," Journal of Nanomaterials, vol. 2011, Article ID 941561, 8 pages, 2011. · ·

23. S. S. Ahluwalia and D. Goyal, "Microbial and plant derived biomass for removal of heavy metals from wastewater," Bioresource Technology, vol. 98, no. 12, pp. 2243–2257, 2007. ·

24. C. Li, W. Jiang, N. Ma et al., "Bioaccumulation of cadmium by growing Zygosaccharomyces rouxii andSaccharomyces cerevisiae," Bioresource Technology, vol. 155, pp. 116–121, 2014. · ·

25. A. M. A. Abdallah, M. A. Abdallah, A. Beltagy, and E. Siam, "Contents of heavy metals in marine algae from Egyptian Red Sea coast," Toxicological and Environmental Chemistry, vol. 88, no. 1, pp. 9–22, 2006. · ·

26. D. R. Lovley, "Dissimilatory metal reduction," Annual Review of Microbiology, vol. 47, pp. 263–290, 1993. · ·

27. N. Durán, "Use of nanoparticles in soil-water bioremediation processes," in Proceedings of the 5th International Symposium (ISMOM ‹08), Pucón, Chile, 2008.

28. D. R. Majumder, "Bioremediation: copper nanoparticles from electronic-waste," International Journal of Engineering Science and Technology, vol. 4, no. 10, pp. 4380–4389, 2014.

29. T. A. Khan, M. Nazir, I. Ali, and A. Kumar, "Removal of Chromium(VI) from aqueous solution using guar gum-nano zinc oxide biocomposite adsorbent," Arabian Journal of Chemistry, 2013. · ·

30. N. Tamilselvan, K. Saurav, and K. Kannabiran, "Biosorption of selected toxic heavy metals using algal species Acanthopora spicefera," Pharmacologyonline, vol. 1, pp. 518–528, 2011.

31. D. E. Salt, M. Blaylock, N. P. B. A. Kumar, et al., "Phytoremediation: a novel strategy for the removal of toxic metals from the environment using plants," Bio/Technology, vol. 13, no. 5, pp. 468–474, 1995. · ·

32. I. Raskin, P. B. A. N. Kumar, S. Dushenkov, and D. E. Salt, "Bioconcentration of heavy metals by plants,"Current Opinion in Biotechnology, vol. 5, no. 3, pp. 285–290, 1994. · ·

33. D. E. Salt, R. D. Smith, and I. Raskin, "Phytoremediation," Annual Review of Plant Physiology and Plant Molecular Biology, vol. 49, pp. 643–668, 1998. · ·

34. W. N. C. Anderson, "Hyperaccumulation by plants," in Element Recovery and Sustainability, A. J. Hunt, Ed., Royal Society of Chemistry, 2013.

35. E. Masarovičová and K. Král'ová, "Plant-heavy metal interaction: Phytoremediation, Biofortification and Nanoparticles," in Advances in Selected Plant Physiology Aspects, G. Montanaro, Ed., In-Tech, Rijeka, Croatia, 2012.

36. A. K. Gupta, S. Dwivedi, S. Sinha, R. D. Tripathi, U. N. Rai, and S. N. Singh, "Metal accumulation and growth performance of Phaseolus vulgaris grown in fly ash amended soil," Bioresource Technology, vol. 98, no. 17, pp. 3404–3407, 2007. · ·

37. M. L. Guerinot and D. E. Salt, "Fortified foods and phytoremediation: two sides of the same coin," Plant Physiology, vol. 125, no. 1, pp. 164–167, 2001. · ·

38. T. J. Beveridge and R. G. E. Murray, "Uptake and retention of metals by cell walls of Bacillus subtilis," Journal of Bacteriology, vol. 127, no. 3, pp. 1502–1518, 1976.

39. T. J. Beveridge and W. S. Fyfe, "Metal fixation by bacterial cell walls," Canadian Journal of Earth Sciences, vol. 22, no. 12, pp. 1893–1898, 1985. · ·

40. T. A. Davis, B. Volesky, and A. Mucci, "A review of the biochemistry of heavy metal biosorption by brown algae," Water Research, vol. 37, no. 18, pp. 4311–4330, 2003. · ·

41. B. Volesky and Z. R. Holan, "Biosorption of heavy metals," Biotechnology Progress, vol. 11, no. 3, pp. 235–250, 1995. · ·

42. T. M. Roane, K. L. Josephson, and I. L. Pepper, "Dual-bioaugmentation strategy to enhance remediation of cocontaminated soil," Applied and Environmental Microbiology, vol. 67, no. 7, pp. 3208–3215, 2001. · ·

43. T. Srinath, T. Verma, P. W. Ramteke, and S. K. Garg, "Chromium (VI) biosorption and bioaccumulation by chromate resistant bacteria," Chemosphere, vol. 48, no. 4, pp. 427–435, 2002. · ·

44. M. Fomina and G. M. Gadd, "Biosorption: current perspectives on concept, definition and application," Bioresource Technology, vol. 160, pp. 3–14, 2014. · ·

45. H. Guha, K. Jayachandran, and F. Maurrasse, "Microbiological reduction of chromium(VI) in presence of pyrolusite-coated sand by Shewanella alga Simidu ATCC 55627 in laboratory column experiments,"Chemosphere, vol. 52, no. 1, pp. 175–183, 2003. · ·

46. Z. Chen, Z. Huang, Y. Cheng et al., "Cr(VI) uptake mechanism of Bacillus cereus," Chemosphere, vol. 87, no. 3, pp. 211–216, 2012. · ·

47. Z. He, F. Gao, T. Sha, Y. Hu, and C. He, "Isolation and characterization of a Cr(VI)-reductionOchrobactrum sp. strain CSCr-3 from chromium landfill," Journal of Hazardous Materials, vol. 163, no. 2-3, pp. 869–873, 2009. · ·

48. A. G. Tekerlekopoulou, M. Tsiflikiotou, L. Akritidou et al., "Modelling of biological Cr(VI) removal in draw-fill reactors using microorganisms in suspended and attached growth systems," Water Research, vol. 47, no. 2, pp. 623–636, 2013. · ·

49. V. Somasundaram, L. Philip, and S. M. Bhallamudi, "Experimental and mathematical modeling studies on Cr(VI) reduction by CRB, SRB and IRB, individually and in combination," Journal of Hazardous Materials, vol. 172, no. 2-3, pp. 606–617, 2009. · ·

50. T. Ahmad, I. A. Wani, N. Manzoor, J. Ahmed, and A. M. Asiri, "Biosynthesis, structural characterization and antimicrobial activity of gold and silver nanoparticles," Colloids and Surfaces B: Biointerfaces, vol. 107, pp. 227–234, 2013. · ·

51. A. R. Binupriya, M. Sathishkumar, K. Vijayaraghavan, and S.-I. Yun, "Bioreduction of trivalent aurum to nano-crystalline gold particles by active and inactive cells and cell-free extract of Aspergillus oryzaevar. viridis," Journal of Hazardous Materials, vol. 177, no. 1–3, pp. 539–545, 2010. · ·

52. A. Mishra, S. K. Tripathy, and S.-I. Yun, "Fungus mediated synthesis of gold nanoparticles and their conjugation with genomic DNA isolated from Escherichia coli and Staphylococcus aureus," Process Biochemistry, vol. 47, no. 5, pp. 701–711, 2012. · ·

53. K. B. Narayanan and N. Sakthivel, "Facile green synthesis of gold nanostructures by NADPH-dependent enzyme from the extract of Sclerotium rolfsii," Colloids and Surfaces A: Physicochemical and Engineering Aspects, vol. 380, no. 1–3, pp. 156–161, 2011.

54. F. Cai, J. Li, J. Sun, and Y. Ji, "Biosynthesis of gold nanoparticles by biosorption using Magnetospirillum gryphiswaldense MSR-1," Chemical Engineering Journal, vol. 175, no. 1, pp. 70–75, 2011.

55. S. A. Aromal and D. Philip, "Benincasa hispida seed mediated green synthesis of gold nanoparticles and its optical nonlinearity," Physica E: Low-Dimensional Systems and Nanostructures, vol. 44, no. 7-8, pp. 1329–1334, 2012. · ·

56. N. Kulkarni and U. Muddapur, "Biosynthesis of metal nanoparticles: a review," Journal of Nanotechnology, vol. 2014, Article ID 510246, 8 pages, 2014. · ·

57. M. Rai and A. Yadav, "Plants as potential synthesiser of precious metal nanoparticles: progress and prospects," IET Nanobiotechnology, vol. 7, no. 3, pp. 117–124, 2013. · ·

58. K. B. Narayanan and N. Sakthivel, "Biological synthesis of metal nanoparticles by microbes," Advances in Colloid and Interface Science, vol. 156, no. 1-2, pp. 1–13, 2010. · ·

59. Y. Wang, X. He, K. Wang, X. Zhang, and W. Tan, "Barbated Skullcup herb extract-mediated biosynthesis of gold nanoparticles and its primary application in electrochemistry," Colloids and Surfaces B: Biointerfaces, vol. 73, no. 1, pp. 75–79, 2009. · ·

60. A. J. Hunt, T. J. Farmer, and J. H. Clark, "Elemental sustainability and the importance of scarce element recovery," in Element Recovery and Sustainability, chapter 1, 2013.

61. K. Pollmann, J. Raff, M. Merroun, K. Fahmy, and S. Selenska-Pobell, "Metal binding by bacteria from uranium mining waste piles and its technological applications," Biotechnology Advances, vol. 24, no. 1, pp. 58–68, 2006. · ·

62. S. Wang and X. Zhao, "On the potential of biological treatment for arsenic contaminated soils and groundwater," Journal of Environmental Management, vol. 90, no. 8, pp. 2367–2376, 2009. · ·

63. D. Purkayastha, U. Mishra, and S. Biswas, "A comprehensive review on Cd(II) removal from aqueous solution," Journal of Water Process Engineering, vol. 2, pp. 105–128, 2014. ·

64. S. Islamoglu, L. Yilmaz, and H. O. Ozbelge, "Development of a precipitation based separation scheme for selective removal and recovery of heavy metals from cadmium rich electroplating

industry effluents,"Separation Science and Technology, vol. 41, no. 15, pp. 3367–3385, 2006. · ·

65.   A. Saeed and M. Iqbal, "Bioremoval of cadmium from aqueous solution by black gram husk (Cicer arientinum)," Water Research, vol. 37, no. 14, pp. 3472–3480, 2003. · ·

66.   J. Shah, M. R. Jan, A. U. Haq Atta Ul, and M. Sadia, "Biosorption of cadmium from aqueous solution using mulberry wood sawdust: equilibrium and kinetic studies," Separation Science and Technology, vol. 46, no. 10, pp. 1631–1637, 2011. · ·

67.   C. P. J. Isaac and A. Sivakumar, "Removal of lead and cadmium ions from water using Annona squamosa shell: kinetic and equilibrium studies," Desalination and Water Treatment, vol. 51, no. 40–42, pp. 7700–7709, 2013. · ·

68.   J.-S. Chang, R. Law, and C.-C. Chang, "Biosorption of lead, copper and cadmium by biomass ofPseudomonas aeruginosa PU21," Water Research, vol. 31, no. 7, pp. 1651–1658, 1997. · ·

69.   H. Ni, Z. Xiong, T. Ye, Z. Zhang, X. Ma, and L. Li, "Biosorption of copper(II) from aqueous solutions using volcanic rock matrix-immobilized Pseudomonas putida cells with surface-displayed cyanobacterial metallothioneins," Chemical Engineering Journal, vol. 204-205, pp. 264–271, 2012. · ·

70.   A. Hammaini, F. González, A. Ballester, M. L. Blázquez, and J. A. Muñoz, "Biosorption of heavy metals by activated sludge and their desorption characteristics," Journal of Environmental Management, vol. 84, no. 4, pp. 419–426, 2007. · ·

71.   W.-B. Lu, J.-J. Shi, C.-H. Wang, and J.-S. Chang, "Biosorption of lead, copper and cadmium by an indigenous isolate Enterobacter sp. J1 possessing high heavy-metal resistance," Journal of Hazardous Materials, vol. 134, no. 1–3, pp. 80–86, 2006. · ·

72.   W.-J. Liu, F.-X. Zeng, H. Jiang, X.-S. Zhang, and H.-Q. Yu, "Techno-economic evaluation of the integrated biosorption-pyrolysis technology for lead (Pb) recovery from aqueous solution," Bioresource Technology, vol. 102, no. 10, pp. 6260–6265, 2011. · ·

73.   K. N. Thakkar, S. S. Mhatre, and R. Y. Parikh, "Biological synthesis of metallic nanoparticles,"Nanomedicine: Nanotechnology, Biology, and Medicine, vol. 6, no. 2, pp. 257–262, 2010. · ·

74. A. K. Mittal, Y. Chisti, and U. C. Banerjee, "Synthesis of metallic nanoparticles using plant extracts,"Biotechnology Advances, vol. 31, no. 2, pp. 346–356, 2013. · ·

75. S. Gupta, K. Sharma, and R. Sharma, "Myconanotechnology and application of nanoparticles in biology," Recent Research in Science and Technology, vol. 4, no. 8, pp. 36–38, 2012.

76. N. Krumov, I. Perner-Nochta, S. Oder, V. Gotcheva, A. Angelov, and C. Posten, "Production of inorganic nanoparticles by microorganisms," Chemical Engineering and Technology, vol. 32, no. 7, pp. 1026–1035, 2009. · ·

77. K. V. Pavani, N. Sunil Kumar, and K. Gayathramma, "Plants as ecofriendly nanofactories," Journal of Bionanoscience, vol. 6, no. 1, pp. 1–6, 2012. · ·

78. S. Baker, B. P. Harini, D. Rakshith, and S. Satish, "Marine microbes: invisible nanofactories," Journal of Pharmacy Research, vol. 6, pp. 383–388, 2013.

79. N. Duran, D. P. Marcato, A. Ingle, A. Gade, and M. Rai, "Fungi-mediated synthesis of silver nanoparticles: characterization processes and applications," in Progress in Mycology, M. Rai and G. Kövics, Eds., pp. 425–449, Scientific Publishers, Jodhpur, India, 2010.

80. S. Baker, D. Rakshith, K. S. Kavitha et al., "Plants: emerging as nanofactories towards facile route in synthesis of nanoparticles," BioImpacts, vol. 3, no. 3, pp. 111–117, 2013. · ·

81. V. Parashar, R. Parashar, B. Sharma, and A. C. Pandey, "Parthenium leaf extract mediated synthesis of silver nanoparticles: a novel approach towards weed utilization," Digest Journal of Nanomaterials and Biostructures, vol. 4, no. 1, pp. 45–50, 2009.

82. K. S. Prasad, H. Patel, T. Patel, K. Patel, and K. Selvaraj, "Biosynthesis of Se nanoparticles and its effect on UV-induced DNA damage," Colloids and Surfaces B: Biointerfaces, vol. 103, pp. 261–266, 2013. · ·

83. X. Weng, L. Huang, Z. Chen, M. Megharaj, and R. Naidu, "Synthesis of iron-based nanoparticles by green tea extract and their degradation of malachite," Industrial Crops and Products, vol. 51, pp. 342–347, 2013. · ·

84. G. Rajakumar, A. A. Rahuman, B. Priyamvada, V. G. Khanna, D. K. Kumar, and P. J. Sujin, "Eclipta prostrata leaf aqueous extract mediated synthesis of titanium dioxide nanoparticles," Materials Letters, vol. 68, pp. 115–117, 2012. · ·

85. G. K. Naik, P. M. Mishra, and K. Parida, "Green synthesis of Au/ TiO_2 for effective dye degradation in aqueous system," Chemical Engineering Journal, vol. 229, pp. 492–497, 2013. · ·

86. B. Zheng, T. Kong, X. Jing et al., "Plant-mediated synthesis of platinum nanoparticles and its bioreductive mechanism," Journal of Colloid and Interface Science, vol. 396, pp. 138–145, 2013. · ·

87. V.Vignesh, K. Felix Anbarasi, S. Karthikeyeni, G. Sathiyanarayanan, P. Subramanian, and R. Thirumurugan, "A superficial phyto-assisted synthesis of silver nanoparticles and their assessment on hematological and biochemical parameters in Labeo rohita (Hamilton, 1822)," Colloids and Surfaces A: Physicochemical and Engineering Aspects, vol. 439, pp. 184–192, 2013. · ·

88. M. S. Abdel-Aziz, M. S. Shaheen, A. A. El-Nekeety, and M. A. Abdel-Wahhab, "Antioxidant and antibacterial activity of silver nanoparticles biosynthesized using Chenopodium murale leaf extract,"Journal of Saudi Chemical Society, vol. 18, no. 4, pp. 356–363, 2014. · ·

89. A. J. Kora, R. B. Sashidhar, and J. Arunachalam, "Aqueous extract of gum olibanum (Boswellia serrata): a reductant and stabilizer for the biosynthesis of antibacterial silver nanoparticles," Process Biochemistry, vol. 47, no. 10, pp. 1516–1520, 2012. · ·

90. V. Gopinath, S. Priyadarshini, N. Meera Priyadharsshini, K. Pandian, and P. Velusamy, "Biogenic synthesis of antibacterial silver chloride nanoparticles using leaf extracts of Cissus quadrangularis Linn,"Materials Letters, vol. 91, pp. 224–227, 2013. · ·

91. Y. Zhang, X. Cheng, Y. Zhang, X. Xue, and Y. Fu, "Biosynthesis of silver nanoparticles at room temperature using aqueous aloe leaf extract and antibacterial properties," Colloids and Surfaces A: Physicochemical and Engineering Aspects, vol. 423, pp. 63–68, 2013. · ·

92. V. Gopinath, D. MubarakAli, S. Priyadarshini, N. M. Priyadharsshini, N. Thajuddin, and P. Velusamy, "Biosynthesis of silver nanoparticles from Tribulus terrestris and its antimicrobial activity: a novel biological approach," Colloids and Surfaces B: Biointerfaces, vol. 96, pp. 69–74, 2012. · ·

93. M. Yilmaz, H. Turkdemir, M. A. Kilic et al., "Biosynthesis of silver nanoparticles using leaves of Stevia rebaudiana," Materials Chemistry and Physics, vol. 130, no. 3, pp. 1195–1202, 2011. · ·

94. M. Vijayakumar, K. Priya, F. T. Nancy, A. Noorlidah, and A. B. A. Ahmed, "Biosynthesis, characterisation and anti-bacterial effect of plant-mediated silver nanoparticles using Artemisia nilagirica," Industrial Crops and Products, vol. 41, no. 1, pp. 235–240, 2013. · ·

95. T. Y. Suman, S. R. Radhika Rajasree, A. Kanchana, and S. B. Elizabeth, "Biosynthesis, characterization and cytotoxic effect of plant mediated silver nanoparticles using Morinda citrifolia root extract," Colloids and Surfaces B: Biointerfaces, vol. 106, pp. 74–78, 2013. · ·

96. J. J. Antony, P. Sivalingam, D. Siva et al., "Comparative evaluation of antibacterial activity of silver nanoparticles synthesized using Rhizophora apiculata and glucose," Colloids and Surfaces B: Biointerfaces, vol. 88, no. 1, pp. 134–140, 2011. · ·

97. K. Raja, A. Saravanakumar, and R. Vijayakumar, "Efficient synthesis of silver nanoparticles from Prosopis juliflora leaf extract and its antimicrobial activity using sewage," Spectrochimica Acta A: Molecular and Biomolecular Spectroscopy, vol. 97, pp. 490–494, 2012. · ·

98. M. M. H. Khalil, E. H. Ismail, K. Z. El-Baghdady, and D. Mohamed, "Green synthesis of silver nanoparticles using olive leaf extract and its antibacterial activity," Arabian Journal of Chemistry, 2013.

99. T. J. I. Edison and M. G. Sethuraman, "Instant green synthesis of silver nanoparticles using Terminalia chebula fruit extract and evaluation of their catalytic activity on reduction of methylene blue," Process Biochemistry, vol. 47, no. 9, pp. 1351–1357, 2012. · ·

100. S. M. Roopan, G. Madhumitha, A. A. Rahuman, C. Kamaraj, A. Bharathi, and T. V. Surendra, "Low-cost and eco-friendly phyto-synthesis of silver nanoparticles using Cocos nucifera coir extract

and its larvicidal activity," Industrial Crops and Products, vol. 43, no. 1, pp. 631–635, 2013. · ·

101. M. F. Zayed, W. H. Eisa, and A. A. Shabaka, "Malva parviflora extract assisted green synthesis of silver nanoparticles," Spectrochimica Acta A: Molecular and Biomolecular Spectroscopy, vol. 98, pp. 423–428, 2012. · ·

102. D. Philip, "Mangifera Indica leaf-assisted biosynthesis of well-dispersed silver nanoparticles," Spectrochimica Acta A: Molecular and Biomolecular Spectroscopy, vol. 78, no. 1, pp. 327–331, 2011. · ·

103. N. Yang and W.-H. Li, "Mango peel extract mediated novel route for synthesis of silver nanoparticles and antibacterial application of silver nanoparticles loaded onto non-woven fabrics," Industrial Crops and Products, vol. 48, pp. 81–88, 2013. · ·

104. R. Sankar, A. Karthik, A. Prabu, S. Karthik, K. S. Shivashangari, and V. Ravikumar, "Origanum vulgare mediated biosynthesis of silver nanoparticles for its antibacterial and anticancer activity," Colloids and Surfaces B: Biointerfaces, vol. 108, pp. 80–84, 2013. · ·

105. K. Mallikarjuna, N. John Sushma, G. Narasimha, L. Manoj, and B. Deva Prasad Raju, "Phytochemical fabrication and characterization of silver nanoparticles by using Pepper leaf broth," Arabian Journal of Chemistry, 2012. · ·

106. R. Arunachalam, S. Dhanasingh, B. Kalimuthu, M. Uthirappan, C. Rose, and A. B. Mandal, "Phytosynthesis of silver nanoparticles using Coccinia grandis leaf extract and its application in the photocatalytic degradation," Colloids and Surfaces B: Biointerfaces, vol. 94, pp. 226–230, 2012. · ·

107. V. S. Kotakadi, Y. S. Rao, S. A. Gaddam, T. N. V. K. V. Prasad, A. V. Reddy, and S. Gopal, "Simple and rapid biosynthesis of stable silver nanoparticles using dried leaves of Catharanthus roseus. Linn. G. Donn and its anti microbial activity," Colloids and Surfaces B: Biointerfaces, vol. 105, pp. 194–198, 2013. · ·

108. N. Basavegowda and Y. R. Lee, "Synthesis of silver nanoparticles using Satsuma mandarin (Citrus unshiu) peel extract: a novel approach towards waste utilization," Materials Letters, vol. 109, pp. 31–33, 2013. · ·

109. C. Tamuly, M. Hazarika, and M. Bordoloi, "Biosynthesis of Au nanoparticles by Gymnocladus assamicusand its catalytic activity," Materials Letters, vol. 108, pp. 276–279, 2013. · ·

110. K. Sneha, M. Sathishkumar, S. Kim, and Y.-S. Yun, "Counter ions and temperature incorporated tailoring of biogenic gold nanoparticles," Process Biochemistry, vol. 45, no. 9, pp. 1450–1458, 2010. · ·

111. L. Castro, M. L. Blázquez, F. González, J. A. Muñoz, and A. Ballester, "Extracellular biosynthesis of gold nanoparticles using sugar beet pulp," Chemical Engineering Journal, vol. 164, no. 1, pp. 92–97, 2010. · ·

112. S. U. Ganaie, T. Abbasi, J. Anuradha, and S. A. Abbasi, "Biomimetic synthesis of silver nanoparticles using the amphibious weed ipomoea and their application in pollution control," Journal of King Saud University, vol. 26, no. 3, pp. 222–229, 2014. · ·

113. P. Lodeiro and M. Sillanpää, "Gold reduction in batch and column experiments using silica gel derivates and seaweed biomass," Chemical Engineering Journal, vol. 230, pp. 372–379, 2013. · ·

114. I. K. Sen, K. Maity, and S. S. Islam, "Green synthesis of gold nanoparticles using a glucan of an edible mushroom and study of catalytic activity," Carbohydrate Polymers, vol. 91, no. 2, pp. 518–528, 2013. · ·

115. K. P. Kumar, W. Paul, and C. P. Sharma, "Green synthesis of gold nanoparticles with Zingiber officinaleextract: characterization and blood compatibility," Process Biochemistry, vol. 46, no. 10, pp. 2007–2013, 2011. · ·

116. R. Vijayakumar, V. Devi, K. Adavallan, and D. Saranya, "Green synthesis and characterization of gold nanoparticles using extract of anti-tumor potent Crocus sativus," Physica E: Low-Dimensional Systems and Nanostructures, vol. 44, no. 3, pp. 665–671, 2011.

117. D. Philip, "Green synthesis of gold and silver nanoparticles using Hibiscus rosa sinensis," Physica E: Low-Dimensional Systems and Nanostructures, vol. 42, no. 5, pp. 1417–1424, 2010. · ·

118. N. Gupta, H. P. Singh, and R. K. Sharma, "Single-pot synthesis: plant mediated gold nanoparticles catalyzed reduction of methylene blue in presence of stannous chloride," Colloids and Surfaces A: Physicochemical and Engineering Aspects, vol. 367, no. 1–3, pp. 102–107, 2010. · ·

119. S. M. Ghoreishi, M. Behpour, and M. Khayatkashani, "Green synthesis of silver and gold nanoparticles using Rosa damascena and its primary application in electrochemistry," Physica E: Low-Dimensional Systems and Nanostructures, vol. 44, no. 1, pp. 97–104, 2011. ··

120. K. Sneha, M. Sathishkumar, J. Mao, I. S. Kwak, and Y.-S. Yun, "Corynebacterium glutamicum-mediated crystallization of silver ions through sorption and reduction processes," Chemical Engineering Journal, vol. 162, no. 3, pp. 989–996, 2010. ··

121. J. P. da Costa, A. V. Girão, J. P. Lourenço, O. C. Monteiro, T. Trindade, and M. C. Costa, "Green synthesis of covellite nanocrystals using biologically generated sulfide: potential for bioremediation systems," Journal of Environmental Management, vol. 128, pp. 226–232, 2013. ··

122. G.-Q. Chen, "Plastics completely synthesized by bacteria: polyhydroxyalkanoates," in Plastics from Bacteria: Natural Functions and Applications, vol. 14 of Microbiology Monographs, pp. 17–37, Springer, Berlin, Germany, 2010. ·

123. P. Rogers, J.-S. Chen, and M. J. Zidwick, "Organic acid and solvent production," in The Prokaryotes, pp. 511–755, Springer, New York, NY, USA, 2006. ·

124. J. O›Keeffe, "Environmental and industrial use of Bacillus subtilis," in Global Post, America›s World News Site, 2014.

125. D. N. Correa-Llantén, S. A. Muñoz-Ibacache, M. E. Castro, P. A. Muñoz, and J. M. Blamey, "Gold nanoparticles synthesized by Geobacillus sp. strain ID17 a thermophilic bacterium isolated from Deception Island, Antarctica," Microbial Cell Factories, vol. 12, no. 1, article 75, 2013. ··

126. C. Malarkodi, S. Rajeshkumar, M. Vanaja, K. Paulkumar, G. Gnanajobitha, and G. Annadurai, "Eco-friendly synthesis and characterization of gold nanoparticles using Klebsiella pneumoniae," Journal of Nanostructure in Chemistry, vol. 3, article 30, 2013. ··

127. L. Du, H. Jiang, X. Liu, and E. Wang, "Biosynthesis of gold nanoparticles assisted by Escherichia coliDH5α and its application on direct electrochemistry of hemoglobin," Electrochemistry Communications, vol. 9, no. 5, pp. 1165–1170, 2007. ··

128. P. Sanpui, S. B. Pandey, S. S. Ghosh, and A. Chattopadhyay, "Green fluorescent protein for in situsynthesis of highly uniform Au nanoparticles and monitoring protein denaturation," Journal of Colloid and Interface Science, vol. 326, no. 1, pp. 129–137, 2008. · ·

129. M. I. Husseiny, M. A. El-Aziz, Y. Badr, and M. A. Mahmoud, "Biosynthesis of gold nanoparticles usingPseudomonas aeruginosa," SpectrochimicaActaA: Molecular and Biomolecular Spectroscopy, vol. 67, no. 3-4, pp. 1003–1006, 2007. · ·

130. S. He, Z. Guo, Y. Zhang, S. Zhang, J. Wang, and N. Gu, "Biosynthesis of gold nanoparticles using the bacteria Rhodopseudomonas capsulata," Materials Letters, vol. 61, no. 18, pp. 3984–3987, 2007. · ·

131. P. Arunkumar, M. Thanalakshmi, P. Kumar, and K. Premkumar, "Micrococcus luteus mediated dual mode synthesis of gold nanoparticles: involvement of extracellular α-amylase and cell wall teichuronic acid," Colloids and Surfaces B: Biointerfaces, vol. 103, pp. 517–522, 2013. · ·

132. A. Malhotra, K. Dolma, N. Kaur, Y. S. Rathore, S. Mayilraj, and A. R. Choudhury, "Biosynthesis of gold and silver nanoparticles using a novel marine strain of Stenotrophomonas," Bioresource Technology, vol. 142, pp. 727–731, 2013. · ·

133. P. J. Fesharaki, P. Nazari, M. Shakibaie et al., "Biosynthesis of selenium nanoparticles using Klebsiella pneumoniae and their recovery by a simple sterilization process," Brazilian Journal of Microbiology, vol. 41, no. 2, pp. 461–466, 2010. · ·

134. A. K. Jha and K. Prasad, "Biosynthesis of metal and oxide nanoparticles using Lactobacilli from yoghurt and probiotic spore tablets," Biotechnology Journal, vol. 5, no. 3, pp. 285–291, 2010. · ·

135. K. Kalimuthu, R. S. Babu, D. Venkataraman, M. Bilal, and S. Gurunathan, "Biosynthesis of silver nanocrystals by Bacillus licheniformis," Colloids and Surfaces B: Biointerfaces, vol. 65, no. 1, pp. 150–153, 2008. · ·

136. S. Sadhasivam, P. Shanmugam, and K. Yun, "Biosynthesis of silver nanoparticles by Streptomyces hygroscopicus and antimicrobial activity against medically important pathogenic

microorganisms,"Colloids and Surfaces B: Biointerfaces, vol. 81, no. 1, pp. 358–362, 2010. · ·

137. P. Sivalingam, J. J. Antony, D. Siva, S. Achiraman, and K. Anbarasu, "Mangrove Streptomyces sp. BDUKAS10 as nanofactory for fabrication of bactericidal silver nanoparticles," Colloids and Surfaces B: Biointerfaces, vol. 98, pp. 12–17, 2012. · ·

138. M. M. Ganesh Babu and P. Gunasekaran, "Production and structural characterization of crystalline silver nanoparticles from Bacillus cereus isolate," Colloids and Surfaces B: Biointerfaces, vol. 74, no. 1, pp. 191–195, 2009. · ·

139. X. Wei, M. Luo, W. Li et al., "Synthesis of silver nanoparticles by solar irradiation of cell-free Bacillus amyloliquefaciens extracts and AgNO$_3$," Bioresource Technology, vol. 103, no. 1, pp. 273–278, 2012. · ·

140. N. Srivastava and M. Mukhopadhyay, "Biosynthesis and structural characterization of selenium nanoparticles mediated by Zooglea ramigera," Powder Technology, vol. 244, pp. 26–29, 2013. · ·

141. T. Wang, L. Yang, B. Zhang, and J. Liu, "Extracellular biosynthesis and transformation of selenium nanoparticles and application in H$_2$O$_2$ biosensor," Colloids and Surfaces B: Biointerfaces, vol. 80, no. 1, pp. 94–102, 2010. · ·

142. Z. H. Dhoondia and H. Chakraborty, "Lactobacillus mediated synthesis of silver oxide nanoparticles,"Nanomaterials and Nanotechnology, vol. 2, no. 1, article 15, 2012.

143. A. Vishnu Kirthi, A. Abdul Rahuman, G. Rajakumar et al., "Biosynthesis of titanium dioxide nanoparticles using bacterium Bacillus subtilis," Materials Letters, vol. 65, no. 17-18, pp. 2745–2747, 2011. · ·

144. R. Usha, E. Prabu, M. Palaniswamy, C. K. Venil, and R. Rajendran, "Synthesis of metal oxide nano particles by streptomyces sp for development of antimicrobial textiles," Global Journal of Biotechnology and Biochemistry, vol. 5, no. 3, pp. 153–160, 2010.

145. V. Ca, S. Hirematha, and M. N. Chandraprabhab, "Green synthesis of ZnO nanoparticles by Calotropis gigantea," International Journal of Current Engineering and Technology, pp. 118–120, 2013.

146. K. Prasad and A. K. Jha, "ZnO nanoparticles: synthesis and adsorption study," Natural Science, vol. 1, pp. 129–135, 2009.

147. E. Selvarajan and V. Mohanasrinivasan, "Biosynthesis and characterization of ZnO nanoparticles usingLactobacillus plantarum VITES07," Materials Letters, vol. 112, pp. 180–182, 2013. · ·

148. C. Jayaseelan, A. A. Rahuman, A. V. Kirthi et al., "Novel microbial route to synthesize ZnO nanoparticles using Aeromonas hydrophila and their activity against pathogenic bacteria and fungi,"Spectrochimica Acta. Part A: Molecular and Biomolecular Spectroscopy, vol. 90, pp. 78–84, 2012. · ·

149. H. J. Bai, Z. M. Zhang, Y. Guo, and G. E. Yang, "Biosynthesis of cadmium sulfide nanoparticles by photosynthetic bacteria Rhodopseudomonas palustris," Colloids and Surfaces B: Biointerfaces, vol. 70, no. 1, pp. 142–146, 2009. · ·

150. N. Vigneshwaran, A. A. Kathe, P. V. Varadarajan, R. P. Nachane, and R. H. Balasubramanya, "Biomimetics of silver nanoparticles by white rot fungus, Phaenerochaete chrysosporium," Colloids and Surfaces B: Biointerfaces, vol. 53, no. 1, pp. 55–59, 2006. · ·

151. R. Sanghi and P. Verma, "Biomimetic synthesis and characterisation of protein capped silver nanoparticles," Bioresource Technology, vol. 100, no. 1, pp. 501–504, 2009. · ·

152. N. Vigneshwaran, N. M. Ashtaputre, P. V. Varadarajan, R. P. Nachane, K. M. Paralikar, and R. H. Balasubramanya, "Biological synthesis of silver nanoparticles using the fungus Aspergillus flavus,"Materials Letters, vol. 61, no. 6, pp. 1413–1418, 2007. ·

153. K. C. Bhainsa and S. F. D›Souza, "Extracellular biosynthesis of silver nanoparticles using the fungusAspergillus fumigatus," Colloids and Surfaces B: Biointerfaces, vol. 47, no. 2, pp. 160–164, 2006. · ·

154. N. S. Shaligram, M. Bule, R. Bhambure et al., "Biosynthesis of silver nanoparticles using aqueous extract from the compactin producing fungal strain," Process Biochemistry, vol. 44, no. 8, pp. 939–943, 2009. · ·

155. D. S. Balaji, S. Basavaraja, R. Deshpande, D. B. Mahesh, B. K. Prabhakar, and A. Venkataraman, "Extracellular biosynthesis of

functionalized silver nanoparticles by strains of Cladosporium cladosporioides fungus," Colloids and Surfaces B: Biointerfaces, vol. 68, no. 1, pp. 88–92, 2009. · ·

156. E. Castro-Longoria, A. R. Vilchis-Nestor, and M. Avalos-Borja, "Biosynthesis of silver, gold and bimetallic nanoparticles using the filamentous fungus Neurospora crassa," Colloids and Surfaces B: Biointerfaces, vol. 83, no. 1, pp. 42–48, 2011. · ·

157. A. Ahmad, P. Mukherjee, S. Senapati et al., "Extracellular biosynthesis of silver nanoparticles using the fungus Fusarium oxysporum," Colloids and Surfaces B: Biointerfaces, vol. 28, no. 4, pp. 313–318, 2003. · ·

158. A. Syed and A. Ahmad, "Extracellular biosynthesis of platinum nanoparticles using the fungus Fusarium oxysporum," Colloids and Surfaces B: Biointerfaces, vol. 97, pp. 27–31, 2012. · ·

159. G. Rajakumar, A. A. Rahuman, S. M. Roopan et al., "Fungus-mediated biosynthesis and characterization of TiO_2 nanoparticles and their activity against pathogenic bacteria," Spectrochimica Acta. Part A: Molecular and Biomolecular Spectroscopy, vol. 91, pp. 23–29, 2012. · ·

160. S. A. Khan and A. Ahmad, "Phase, size and shape transformation by fungal biotransformation of bulk TiO_2," Chemical Engineering Journal, vol. 230, pp. 367–371, 2013. · ·

161. A. Mashrai, H. Khanam, and R. N. Aljawfi, "Biological synthesis of ZnO nanoparticles using C. albicansand studying their catalytic performance in the synthesis of steroidal pyrazolines," Arabian Journal of Chemistry, 2013. · ·

162. K. B. Narayanan and N. Sakthivel, "Synthesis and characterization of nano-gold composite using Cylindrocladium floridanum and its heterogeneous catalysis in the degradation of 4-nitrophenol,"Journal of Hazardous Materials, vol. 189, no. 1-2, pp. 519–525, 2011. · ·

163. F. Arockiya Aarthi Rajathi, C. Parthiban, V. Ganesh Kumar, and P. Anantharaman, "Biosynthesis of antibacterial gold nanoparticles using brown alga, Stoechospermum marginatum (kützing),"Spectrochimica Acta. Part A: Molecular and Biomolecular Spectroscopy, vol. 99, pp. 166–173, 2012. · ·

164. K. Sen, P. Sinha, and S. Lahiri, "Time dependent formation of gold nanoparticles in yeast cells: a comparative study," Biochemical Engineering Journal, vol. 55, no. 1, pp. 1–6, 2011. · ·

165. R. Selvakumar, N. Arul Jothi, V. Jayavignesh et al., "As(V) removal using carbonized yeast cells containing silver nanoparticles," Water Research, vol. 45, no. 2, pp. 583–592, 2011. · ·

166. N. Rascio and F. Navari-Izzo, "Heavy metal hyperaccumulating plants: how and why do they do it? And what makes them so interesting?" Plant Science, vol. 180, no. 2, pp. 169–181, 2011. ·

167. S. Sagner, R. Kneer, G. Wanner, J. P. Cosson, B. Deus-Neumann, and M. H. Zenk, "Hyperaccumulation, complexation and distribution of nickel in Sebertia acuminata," Phytochemistry, vol. 47, no. 3, pp. 339–347, 1998. · ·

168. R. S. Boyd, "The defense hypothesis of elemental hyperaccumulation: Status, challenges and new directions," Plant and Soil, vol. 293, no. 1-2, pp. 153–176, 2007. · ·

169. R. R. Brooks, J. Lee, R. D. Reeves, and T. Jaffre, "Detection of nickeliferous rocks by analysis of herbarium specimens of indicator plants," Journal of Geochemical Exploration, vol. 7, pp. 49–57, 1977. · ·

170. R. Chaney, "Plant uptake of inorganic waste," in Land Treatment of Hazardous Wastes, pp. 50–76, 1983.

171. R. D. Reeves, "Hyperaccumulation of trace elements by plant," in Phytoremediation of Metal-Contaminated Soils, J. L. Morel, G. Echevarria, and N. Goncharova, Eds., vol. 68 of Nato Science Series IV: Earth and Environmental Science, pp. 68–25, Springer, New York, NY, USA, 2006.

172. S. D. Cunningham and W. R. Berti, "Remediation of contaminated soils with green plants: an overview,"In Vitro Cellular & Developmental Biology, vol. 29, no. 4, pp. 207–212, 1993. · ·

173. R. R. Brooks, M. F. Chambers, L. J. Nicks, and B. H. Robinson, "Phytomining," Trends in Plant Science, vol. 3, no. 9, pp. 359–362, 1998. · ·

174. S. Wei, Y. Li, J. Zhan, S. Wang, and J. Zhu, "Tolerant mechanisms of Rorippa globosa (Turcz.) Thell. hyperaccumulating Cd explored from root morphology," Bioresource Technology, vol. 118, pp. 455–459, 2012. · ·

175. A. Baker and R. Brooks, "Terrestrial higher plants which hyperaccumulate metallic elements-a review of their distribution, ecology and phytochemistry," Biorecovery, vol. 1, pp. 81–126, 1989.

176. R. Chaney, M. Mallik, Y. Li, S. Brown, and E. Brewer, "Phytoremediation of soil metals," Current Opinion in Biotechnology, vol. 8, no. 3, pp. 279–284, 1997. ·

177. L. Q. Ma, K. M. Komar, C. Tu, W. H. Zhang, Y. Cai, and E. D. Kennelley, "A fern that hyperaccumulates arsenic," Nature, vol. 409, article 579, 2001.

178. S. Wei, Q. Zhou, X. Wang, K. Zhang, G. Guo, and L. Q. Ma, "A newly-discovered Cd-hyperaccumulatorSolanum nigrum L," Chinese Science Bulletin, vol. 50, no. 1, pp. 33–38, 2005. · ·

179. T. Jaffre, R. R. Brooks, J. Lee, and R. D. Reeves, "Sebertiaacuminata: a nickel accumulating plant from New Caledonia," Science, vol. 193, p. 579, 1976.

180. A. J. M. Baker, R. D. Reeves, and A. S. M. Hajar, "Heavy metal accumulation and tolerance in British populations of the metallophyte Thlaspi caerulescens J & C Presl (Brassicaceae)," The New Phytologist, vol. 127, no. 1, pp. 61–68, 1994.

181. R. D. Reeves and R. R. Brooks, "Hyperaccumulation of lead and zinc by two metallophytes from mining areas of central Europe," Environmental Pollution Series A: Ecological and Biological, vol. 31, no. 4, pp. 277–285, 1983. · ·

182. R. L. Chaney, J. S. Angle, C. L. Broadhurst, C. A. Peters, R. V. Tappero, and D. L. Sparks, "Improved understanding of hyperaccumulation yields commercial phytoextraction and phytomining technologies," Journal of Environmental Quality, vol. 36, no. 5, pp. 1429–1433, 2007. · ·

183. A. van der Ent, A. J. M. Baker, R. D. Reeves, A. J. Pollard, and H. Schat, "Hyperaccumulators of metal and metalloid trace elements: Facts and fiction," Plant and Soil, vol. 362, no. 1-2, pp. 319–334, 2013. · ·

184. L. J. Nicks and M. F. Chambers, "Farming for metals," Mining Environmental Management, vol. 3, pp. 15–18, 1995.

185. R. D. Reeves, R. R. Brooks, and R. M. McFarlane, "Nickel uptake by Californian Streptanthus andCaulanthus with particular

reference to the hyperaccumulator S. polygaloides Gray (Brassicaceae),"American Journal of Botany, vol. 68, pp. 708–712, 1981.

186. B. H. Robinson, R. R. Brooks, A. W. Howes, J. H. Kirkman, and P. E. H. Gregg, "The potential of the high-biomass nickel hyperaccumulator Berkheya coddii for phytoremediation and phytomining,"Journal of Geochemical Exploration, vol. 60, no. 2, pp. 115–126, 1997. · ·

187. C. W. N. Anderson, R. B. Stewart, F. N. Moreno, et al., "Gold phytomining. Novel developments in a plant-based mining system," in Proceedings of the Gold 2003 Conference : New Industrial Applications of Gold, World Gold Council and Canadian Institute of Mining, Metallurgy and Petroleum, 2003.

188. R. D. Reeves and A. J. M. Baker, "Studies on metal uptake by plants from serpentine and non-serpentine populations of Thlaspi goesingense Halacsy (Cruciferae)," New Phytologist, vol. 98, no. 1, pp. 191–204, 1984. · ·

189. B. H. Robinson, R. R. Brooks, and B. E. Clothier, "Soil amendments affecting nickel and cobalt uptake byBerkheya coddii: Potential use for phytomining and phytoremediation," Annals of Botany, vol. 84, no. 6, pp. 689–694, 1999. · ·

190. J. Lin, W. Jiang, and D. Liu, "Accumulation of copper by roots, hypocotyls, cotyledons and leaves of sunflower (Helianthus annuus L.)," Bioresource Technology, vol. 86, no. 2, pp. 151–155, 2003. · ·

191. W. Jiang, D. Liu, and W. Hou, "Hyperaccumulation of cadmium by roots, bulbs and shoots of garlic (Allium sativum L.)," Bioresource Technology, vol. 76, no. 1, pp. 9–13, 2001. · ·

192. S. Wei, Q. Zhou, J. Zhan et al., "Poultry manured Bidens tripartite L. extracting Cd from soil—potential for phytoremediating Cd contaminated soil," Bioresource Technology, vol. 101, no. 22, pp. 8907–8910, 2010. · ·

193. C. W. N. Anderson, R. R. Brooks, A. Chiarucci et al., "Phytomining for nickel, thallium and gold,"Journal of Geochemical Exploration, vol. 67, no. 1–3, pp. 407–415, 1999. · ·

194. F. A. Msuya, R. R. Brooks, and C. W. N. Anderson, "Chemically-induced uptake of gold by root crops: its significance for phytomining," Gold Bulletin, vol. 33, no. 4, pp. 134–137, 2000.

195. T. Jaffré, Etude Écologique du Peuplementvégétal des Sols Dérivés de Rochesultrabasiques en Nouvelle-Calédonie, vol. 124 of Travaux et Documents de l>ORSTOM, ORSTOM, 1980.

196. M. Leblanc, D. Petit, A. Deram, B. H. Robinson, and R. R. Brooks, "The phytomining and environmental significance of hyperaccumulation of thallium by Iberis intermedia from southern France," Economic Geology, vol. 94, no. 1, pp. 109–114, 1999. ·

197. L. F. Juárez-Santillán, C. A. Lucho-Constantino, G. A. Vázquez-Rodríguez, N. M. Cerón-Ubilla, and R. I. Beltrán-Hernández, "Manganese accumulation in plants of the mining zone of Hidalgo, Mexico,"Bioresource Technology, vol. 101, no. 15, pp. 5836–5841, 2010. · ·

198. B. J. Alloway, Heavy Metals, Blackie and Professional, London, UK, 1995.

199. C. W. N. Anderson, R. R. Brooks, R. B. Stewart, and R. Simcock, "Harvesting a crop of gold in plants,"Nature, vol. 395, no. 6702, pp. 553–554, 1998. · ·

200. C. W. N. Anderson, R. R. Brooks, R. B. Stewart, and R. Simcock, "Gold uptake by plants," Gold Bulletin, vol. 32, no. 2, pp. 48–58, 1999. · ·

201. M. Sakakibara, A. Harada, S. Sano, R. S. Hori, and M. Inouhe, "Phytoremediation of heavy metals contaminated by Eleocharis acicularis," in Proceedings of the 12th Symposium Soil Groundwater Contamination Remedies, pp. 545–548, Kyoto, Japan, 2006.

202. N. T. H. Ha, M. Sakakibara, S. Sano, R. S. Hori, and K. Sera, "The potential of Eleocharis acicularis for phytoremediation: case study at an abandoned mine site," Clean, vol. 37, no. 3, pp. 203–208, 2009. · ·

203. N. T. H. Ha, M. Sakakibara, and S. Sano, "Accumulation of Indium and other heavy metals by Eleocharis acicularis: an option for phytoremediation and phytomining," Bioresource Technology, vol. 102, no. 3, pp. 2228–2234, 2011. · ·

204. E. Chmielewská and J. Medved, "Bioaccumulation of heavy metals by green algae Cladophora glomeratain a refinery sewage lagoon," Croatica Chemica Acta, vol. 74, no. 1, pp. 135–145, 2001.

205. D. Kratochvil and B. Volesky, "Advances in the biosorption of heavy metals," Trends in Biotechnology, vol. 16, no. 7, pp. 291–300, 1998. · ·

206. P. Kujan, A. Prell, H. Šafář, M. Sobotka, T. Řezanka, and P. Holler, "Removal of copper ions from dilute solutions by Streptomyces noursei mycelium. Comparison with yeast biomass," Folia Microbiologica, vol. 50, no. 4, pp. 309–313, 2005. · ·

207. D. Brady and J. R. Duncan, "Bioaccumulation of metal cations by Saccharomyces cerevisiae," Applied Microbiology and Biotechnology, vol. 41, no. 1, pp. 149–154, 1994. · ·

208. N. R. Bishnoi and Garima, "Fungus—an alternative for bioremediation of heavy metal containing wastewater: a review," Journal of Scientific and Industrial Research, vol. 64, no. 2, pp. 93–100, 2005.

209. N. P. Bhatia, A. E. Nkang, K. B. Walsh, A. J. M. Baker, N. Ashwath, and D. J. Midmore, "Successful seed germination of the nickel hyperaccumulator Stackhousia tryonii," Annals of Botany, vol. 96, no. 1, pp. 159–163, 2005. · ·

210. R. D. Reeves, A. J. M. Baker, A. Borhidi, and R. Berazaín, "Nickel hyperaccumulation in the serpentine flora of Cuba," Annals of Botany, vol. 83, no. 1, pp. 29–38, 1999. · ·

211. R. D. Reeves and A. J. M. Baker, "Metal accumulating plants," in Phytoremediation of Toxic Metals: Using Plants to Clean up the Environment, I. Raskin and E. D. Ensley, Eds., pp. 193–229, John Wiley & Sons, New York, NY, USA, 2000.

212. J. L. Gardea-Torresdey, J. R. Peralta-Videa, G. de la Rosa, and J. G. Parsons, "Phytoremediation of heavy metals and study of the metal coordination by X-ray absorption spectroscopy," Coordination Chemistry Reviews, vol. 249, no. 17-18, pp. 1797–1810, 2005. · ·

213. A. T. Harris and R. Bali, "On the formation and extent of uptake of silver nanoparticles by live plants,"Journal of Nanoparticle Research, vol. 10, no. 4, pp. 691–695, 2008. · ·

214. V. Sheoran, A. S. Sheoran, and P. Poonia, "Phytomining: a review," Minerals Engineering, vol. 22, no. 12, pp. 1007–1019, 2009. · ·

215. V. Sheoran, S. Sheoran, and A. P. Poonia, "Phytomining of gold: a review," Journal of Geochemical Exploration, vol. 128, pp. 42–50, 2013. · ·

216. V. Wilson-Corral, C. W. N. Anderson, and M. Rodriguez-Lopez, "Gold phytomining: a review of the relevance of this technology to mineral extraction in the 21st century," Journal of Environmental Management, vol. 111, pp. 249–257, 2012. · ·

217. S. Mann, Biomineralization, Oxford University Press, Oxford, UK, 2002.

218. I. C. Gebeshuber, "Biomineralization in marine organisms: status, challenges and prospects for biotechnology," in Springer Handbook of Marine Biotechnology, S.-K. Kim, Ed., pp. 1283–1304, Springer, New York, NY, USA, 2014.

219. G. A. Mansoori, T. F. George, G. Zhang, and L. Assoufid, "Molecular building blocks for nanotechnology," in Molecular Building Blocks for Nanotechnology: From Diamondoids to Nanoscale Materials and Applications, vol. 111, Springer, New York, NY, USA, 2007.

220. V. K. Sharma, R. A. Yngard, and Y. Lin, "Silver nanoparticles: green synthesis and their antimicrobial activities," Advances in Colloid and Interface Science, vol. 145, no. 1-2, pp. 83–96, 2009. · ·

221. M. Zhou, Z. Wei, H. Qiao, L. Zhu, H. Yang, and T. Xia, "Particle size and pore structure characterization of silver nanoparticles prepared by confined arc plasma," Journal of Nanomaterials, vol. 2009, Article ID 968058, 5 pages, 2009. · ·

222. S. R. Waghmare, M. N. Mulla, S. R. Marathe, and K. D. Sonawane, "Ecofriendly production of silver nanoparticles using Candida utilis and its mechanistic action against pathogenic microorganisms," 3 Biotech, 2014. ·

223. Q. H. Tran, V. Q. Nguyen, and A.-T. Le, "Silver nanoparticles: synthesis, properties, toxicology, applications and perspectives," Advances in Natural Sciences: Nanoscience and Nanotechnology, vol. 4, Article ID 033001, 2013. · ·

224. A. Ravindran, P. Chandran, and S. S. Khan, "Biofunctionalized silver nanoparticles: advances and prospects," Colloids and Surfaces B: Biointerfaces, vol. 105, pp. 342–352, 2013. · ·

4

Development of a Bioremediation Technology for the Removal of Thiocyanate from Aqueous Industrial Wastes Using Metabolically Active Microorganisms

Yogesh B. Patil[1]

[1]Symbiosis Institute of Research and Innovation (SIRI), Symbiosis International University (SIU), Lavale, Pune, Maharashtra, India

INTRODUCTION

Contamination of water and soil environment due to the release of toxic and hazardous chemicals as a result of industrialization has taken its toll by causing environmental pollution. If not treated and managed appropriately, toxic and hazardous pollutants may cause severe detrimental (negative), reversible or irreversible, intangible and incapacitating impacts on all forms of living cells. Thiocyanate ($N\equiv C-S^-$)

is one such known hazardous chemical and an important member of cyanide (CN$^-$) family. It is a simple inorganic and one carbon (C-1) compound. Despite its toxicity, it is introduced into the environment by natural (principally by biological cyanide detoxification processes) as well as industrial processes (Kelly and Baker, 1990; Wood, 1975). Thiocyanate (SCN$^-$) has some novel properties. It is linear in nature, electronegative polyatomic ion and a good example of pseudo halide; and therefore produced on a grand scale for its use in diverse industrial processes such as dyeing, acrylic fiber production, thiourea production, photo-finishing, herbicide and insecticide production, metal extraction and electroplating industries (Hughes, 1975). SCN$^-$ is also known for its applications in soil sterilization and corrosion inhibition (Beekhuis, 1975). Consequently, these industries emanate large volumes of SCN$^-$ bearing wastewaters. Apart from SCN$^-$, these effluents might contain other contaminants like heavy metals, cyanide (CN$^-$), metal-cyanides (M$_x$CN) and metal-thiocyanates (M$_x$SCN). Cyanide has the potential to reacts readily with sulphur to produce SCN$^-$ and any industry with cyanide in its waste is a potential source of SCN$^-$ contamination. Steel manufacturing, metal mining and electroplating units are some examples of such industries.

All the species of cyanide family (viz. thiocyanate, cyanide and their metal complexes) have potential to interacting with living cells and strong tendencies to connect to proteins and thereby acts as a non-competitive inhibitor (Westley, 1981). This fact necessitate the industries using and/or emanating SCN$^-$ to adequately detoxify the effluents on priority basis before its discharge in soil and water environment; as it may pose detrimental implications on aquatic life. Moreover, in water scarce situations such untreated and partially treated wastewaters could not be recycled back into the industrial processes. The concentration of SCN$^-$ arising from all the above mentioned sources is normally in the range of 5 - 110 mg/l (Mudder and Whitlock, 1984). Although many statutory agencies across the world have set the statutory limits for cyanide and heavy metal discharge, till date there are no such prescribed limits set or documented for SCN$^-$ discharge. Earlier scientific studies indicate that in general, SCN$^-$ is approximately 7 to 10 times less toxic then free cyanide species. The US-health service cites 0.01 mg/l as guideline and 0.2 mg/l as the permissible limit for cyanide species. In India, the Central Pollution Control Board (CPCB) had set a Minimum National Standard (MINAS) limit for cyanide as 0.2 mg/l. Therefore, the cyanide bearing effluents generated from industries needs suitable treatment to bring down the total cyanide levels below 0.2 mg/l. taking into consideration the mentioned facts, standards for discharge of SCN⁻ could readily be deduced to 1 mg/l to be on the safer side. In order to minimize the risk of exposure to the public and aquatic ecosystems, the clean-up of SCN⁻ contaminated wastewaters is therefore necessary. Patil and Kulkarni (2008) have reported the environmental sensitivity and safety aspects in mining industries in regard to cyanide. Impact of cyanide species on fresh water fish Catla catla have also been reported (Prashanth and Patil, 2008).

Numerous technologies are currently employed to detoxify SCN⁻ bearing effluents; and the most widely being used is direct alkaline chlorination or addition of hypochlorite. However, this method produces large aggregates of chemical sludge, which does not have any further utilization and is environmentally hazardous to handle (Lanza and Bertazzoli, 2002). As per Indian regulations, such hazardous chemical sludge is transited from the industrial location to a specially designed Treatment, Storage and Disposal Facility (TSDF) thereby increasing the overall energy consumption, transportation cost and air pollution. Secondly, chlorination also fails to bring the concentration of SCN⁻ (and other CN⁻ species) within the statutory limits especially when heavy metals are present in the effluents. Thirdly, chlorination increases the total dissolved solids (TDS) content of the treated wastewaters, which makes it unfit for further use. Other physico-chemical processes like hydrogen peroxide oxidation, ozone oxidation, electrolytic decomposition, etc. are highly expensive and are rarely used for the treatment of SCN⁻. Thus, there is a pressing need for the development of an alternative treatment process capable of achieving high degradation efficiency at low cost.

Bioremediation (biological treatment system) using metabolically active (live) microorganism is one such effective alternative for the detoxification of toxic chemical wastes. This process has immense potential of treating variety of pollutants (both toxic and non-toxic); has several advantages over conventional methods and therefore being explored by the researchers world-wide. Microorganisms capable of utilizing C-1 compounds like CN⁻ and M$_x$CNs are well documented and have been studied for long time (Dash et al., 2009; Gurbuz, et al., 2004; Karavaiko et al., 2000; Patil and Paknikar, 2000a; Patil and Paknikar, 2000b; Patil and Paknikar, 2001; Patil et al., 2012). Some research papers on biodegradation of SCN⁻ have also been reported (Chaudhari and Kodam, 2010; Hung and Pavlostathis, 1998; Patil, 2008a; Van Zyl et al., 2011). Use of metabolically passive (dead or inactive) microorganisms for the removal and recovery of metal-cyanides and SCN⁻ have also been reported (Gaddi and Patil, 2011; Patil, 2012; Patil and Paknikar, 1999; Thakur and Patil, 2009). Successful efforts to set-up large scale bioremediation technology for the treatment of cyanide, metal-cyanide and SCN⁻ from mining effluent have been made on commercial scale (Mudder and Whitlock, 1984). However, there are very few reports on the microbial SCN⁻ degradation from the process development point of view (Patil, 2006; Patil, 2008a; Patil, 2008b; Patil, 2011; Sorokin et al., 2001; Stratford et al., 1994). Moreover, utilization of SCN⁻ by microbes as a suitable growth substrate (carbon and/or nitrogen source) is poorly understood. Lack of scientific knowledge in this regard may pose problems in the biological treatment systems. The author in the present research chapter focuses on the development of a bioremediation technology for the removal of SCN⁻ from aqueous industrial wastes using metabolically active microorganisms.

MATERIALS AND METHODS

Analyses, Chemicals and Glassware

Potassium SCN⁻ (KSCN) was obtained from Qualigens, Mumbai, India. SCN⁻ assay was carried out spectrophotometrically (Spectronic, Model-20D, India) using ferric nitrate method at 460 nm as described in Standard Methods (APHA-AWWA-WEF, 1998). Digital pH meter (Elico, Model Ll-120, and India) was employed to determine pH of solutions. Bacterial population from culture media, activated sludge and soil were determined microscopically (Metzer, India, METZ-778A) using Neubauer's chamber (Fein-Optik, Blankenburg, GDR) and by total viable count (TVC). Analytical grade chemicals were used for all experiments. Reagents were prepared in glass distilled water and stored under refrigerated conditions (8-10°C).

Enrichment and Isolation of SCN⁻ Degrading Bacterial Consortium

Enrichment culture and growth of mixed bacterial community (bacterial consortium) was carried out using M-9 minimal salts medium (MSM) (Patil and Paknikar, 2000a). One litre of medium contained $Na_2HPO_4.2H_2O$ - 3.0 g; KH_2PO_4 - 1.5 g; NaCl - 0.25 g; distilled water - 1000 ml and 1 ml/l trace metal solution (Bauchop and Elsden, 1960; Millar, 1972). The medium pH was adjusted to 7.5 using 1 M NaOH/HCl. Glucose (10 mM) was added as the sole source of carbon and energy. SCN⁻ (50 mg/l) was supplemented to the enrichment medium as the sole source of nitrogen. Enrichment culture for the isolation of SCN⁻ degrading microorganisms were set-up in aerobic and unsterilized conditions using activated sludge (obtained from secondary treatment of sewage treatment plant) and garden soil. Both the samples were collected in clean polythene bags and carried to the laboratory. Two cylindrical glass jars (reactors) of 1200 ml capacity each were employed for the enrichment purpose. Working volume of the glass reactor was 1000 ml. 100 ml of activated sludge or 100 g of garden soil was added in 900 ml M-9 MSM containing SCN⁻ and glucose as the source of nitrogen and carbon, respectively in order to obtain the final concentration of SCN⁻ and glucose as 50 mg/l and 10 mM, respectively. Enrichment was carried out at the pH 7.5. Both the glass reactors were incubated at room temperature (30±2°C). Air was sparged continuously at the bottom of medium at the rate of 1000±50 ml/min using electrical aerator units. Seven to eight successive transfers of 10% solution enriched with bacterial flora were given periodically in the fresh M-9 MSM containing SCN⁻ and glucose as mentioned earlier (Patil, 2006; Patil, 2008a).

Purification and Identification of Bacterial Cultures

The enrichment cultures as obtained were streaked on nutrient agar medium and M-9 agar medium (containing SCN⁻ and glucose) plates in aseptic conditions at 35°C for 48-96 h. In all, six diverse types of bacterial colonies (three each from garden soil and activated sludge enrichment) appeared on the plates. The cultures were further purified and then transferred to nutrient agar and M-9 agar slopes. By way of periodic transfers, one set of bacterial consortium was consistently maintained in liquid medium (i.e. M-9 MSM) (Patil 2008a; Patil, 2008b). Further, the isolated bacterial cultures were subjected for microscopic examination (Gram staining and motility) and cultural characteristics on the nutrient agar plates. Bergey's Manual of Systematic Bacteriology (Holt, 1989) was used to identify the SCN⁻ degrading bacterial cultures up to genus level only.

SCN⁻ Degradation Efficiency of the Isolated Bacterial Cultures

Quantitative studies on SCN⁻ degradation were performed to determine the efficiency of isolated cultures in their individual capacity and mixed form. Experiments were performed in 250 ml Erlenmeyer flasks containing 100 ml sterile M-9 MSM (pH 7.0) and 50 mg/L potassium thiocyanate (KSCN), which acted as a nitrogen source. 10 mM glucose was used as carbon source. Bacterial cell suspension of 0.1 ml containing 10^8 cells/ml were inoculated into the flasks and were incubated at 30°C in a rotary shaker incubator (Remi, India) at 150 rpm for 48 h. Requisite controls were used and experiments were performed in duplicates and repeated twice. SCN⁻ contents were determined periodically as mentioned earlier. SCN⁻ degradation efficiency of individual and mixed bacterial culture was expressed in terms of percent total SCN⁻ degraded in 48 h. Reaction rate and first order rate constant for SCN⁻ biodegradation were calculated experimentally using equation 1 and 2, respectively (Sellers, 1999).

$$\Delta C / \Delta t = k\, C \tag{1}$$

$$\ln C_t - \ln C_o = - k\, t\, \mu \tag{2}$$

Where, C = SCN⁻ concentration (mg/l); t = time (h); k = rate of reaction / first order rate constant (h^{-1}); C_o = initial SCN⁻ concentration (mg/l) and C_t = SCN⁻ concentration at time t.

Utilization of SCN⁻ as the Sole Source of Cellular Nitrogen by Bacterial Consortium

Batch mode experiments on SCN⁻ biodegradation were conducted in aseptic conditions with 100 ml M-9 MSM in 250 ml conical flasks. The medium was augmented with carbon and nitrogen sources with different combinations as mentioned: - (i) potassium thiocyanate - KSCN (50 mg/l) as the sole carbon and nitrogen source or (ii) KSCN (50 mg/l) and glucose (10 mM) as the sole nitrogen and carbon, respectively or (iii) KSCN (50 mg/l) and NH₄Cl (1 mM) as the sole carbon and nitrogen, respectively or (iv) KSCN (50 mg/l), NH₄Cl (1 mM) and glucose (10 mM) as the nitrogen, nitrogen and carbon sources, respectively. All experiments were conducted at pH 7.0. Flasks were incubated at 30°C on a rotary shaker incubator (Remi, CIS-24 BL) at 150 rpm for 48-72 h (Patil, 2011).

Factors Influencing SCN⁻ Biodegradation

Series of batch culture experiments were conducted to investigate the influence of various parameters on SCN⁻ biodegradation. Experimental conditions used were as follows: 150 ml capacity Erlenmeyer flasks with 25 ml M-9 MSM containing SCN⁻ (50 mg/l) and glucose (10 mM). 1 ml of previously grown culture (for 24 h) having cell density of 10^8 cells/ml was used as inoculum. The flasks were incubated in stationary conditions. Impact of pH (5.0 - 9.0), temperature (20 – 45 °C), initial cell density (10^5 – 10^9 cells/ml) and glucose (1-20 mM) were checked by running different set of experiments, wherein, one parameter was varied keeping the others constant. Periodic analyses were conducted as mentioned earlier.

Biosorption of SCN⁻ by Bacterial Consortium at High Cell Density

Experiments were performed in 150 ml flasks. 50 ml aliquots of SCN⁻ (50 mg/l) adjusted to optimum pH (7.0) was contacted with bacterial consortium (10^8 cells/ml). The culture was inactivated by boiling for the period of 10 min prior to contact. The flasks were incubated at 30°C on a rotary shaker (150 rpm) for 24 h. Bacterial cells were separated by centrifugation at 1000 rpm for 10 min and the cell free supernatant was subjected to determine the residual SCN⁻ concentration.

Impact of Cations and Anions on SCN⁻ Biodegradation

Batch experiments were performed under optimized conditions as described earlier. Impact of diverse cations, especially heavy metals (0.1 mM each) on biodegradation of SCN⁻ was studied. Metals were added as sulfate salts (range 0.1-1 mM). To study the influence of sulfates, additional sulfate was added to the medium as sodium sulfate. Chlorides were added as sodium chloride in the range of 1-10 mM, while cyanide was added as sodium cyanide (0.1-1 mM).

Degradation of SCN⁻ from Industrial Effluent by Bacterial Consortium

SCN⁻ effluent was synthetically prepared in the laboratory because of the difficulty in procurement of effluent from industry. This was to test the practical applicability of the microbial process for degradation of SCN⁻. Batch experiments were performed as mentioned earlier under optimized conditions (pH 7.0, temperature 30°C and bacterial cell density of 10^8 cells/ml). Thiocyanate served as nitrogen source, while sucrose (COD 500 mg/l) was used as carbon source. Parameters such as pH, SCN⁻, COD and soluble metal content were measured at regular intervals for a period of 48 h.

Treatment of SCN⁻ Waste in a Continuous Treatment System (CTS)

Thiocyanate containing simulated was treated in a continuous treatment system (CTS) as shown in Fig. 1. The CTS comprised of cylindrical glass column (height, 24 cm; diameter, 8 cm and total volume 0.2 L) containing one-litre simulated SCN⁻ effluent (50 mg/l SCN⁻) having COD of 500-600 mg/l. The consortium culture was inoculated at the level of 10^8 cells/ml (final cell density) and the contents of the reactor were stirred by sparging air at the rate of 1000 ml/min. The pH of wastewater supplemented with nutrients was adjusted between 7.0-7.3 (using 1 M NaOH/H$_3$PO$_4$) and then added from the top of the reactor by manual adjustment at the flow rate of approximate 40-50 ml/h as calculated from mass balance equation. The treated effluent was removed from the bottom at the same flow rate. The CTS was operated at ambient temperature (30±2°C) in continuous mode for over a period of 30 days (720 h) by periodically checking the influent and effluent water characteristics for pH, SCN⁻, COD and cell count according to the method prescribed in Standard Methods (APHA-AWWA-WEF, 1998).

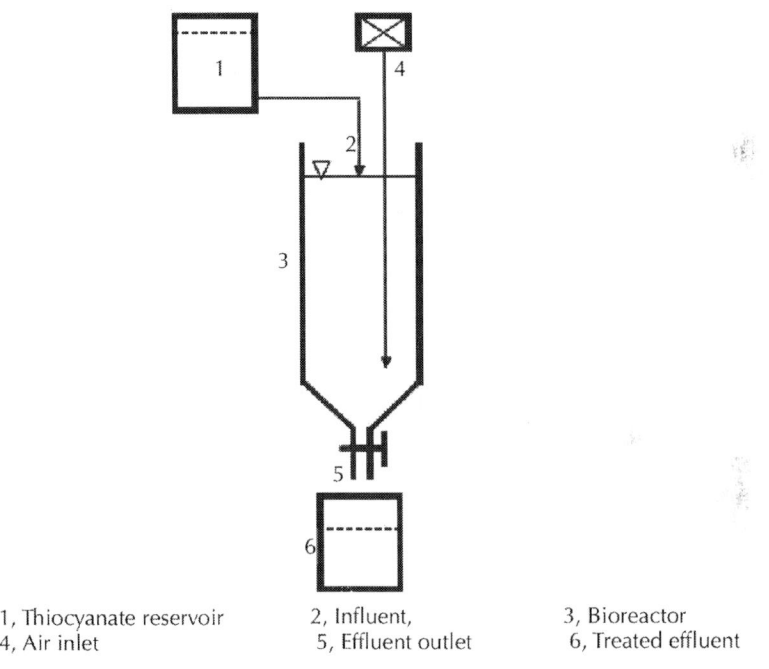

| 1, Thiocyanate reservoir | 2, Influent, | 3, Bioreactor |
| 4, Air inlet | 5, Effluent outlet | 6, Treated effluent |

Figure 1: Schematic outline of laboratory scale Continuous Treatment System (CTS) for degradation of thiocyanate.

RESULTS AND DISCUSSION

Enrichment and Isolation of SCN⁻ Degrading Bacterial Consortium

Both the enrichment culture (garden soil and activated sludge) elucidated that the time incurred reduced significantly with each subsequent transfer cycle for complete disappearance of SCN⁻. Time taken for biodegradation in first, third and fifth cycle was 100, 80 and 70 hours, respectively. After seventh cycle the time taken for biodegradation of >98% SCN⁻ was stabilized around 40-45 hours. Each subsequent transfer was given in fresh M-9 MSM soon after the SCN⁻ concentration reached to < 1 mg/l (efficiency ≥98%) in the previous cycle. The bacterial count was consistently $>10^8$ cells/ml during each transfer cycle. pH and total viable count (TVC) of garden soil prior to enrichment was 8.12 and 3.5×10^8 cells/ml, respectively; and for activated sludge it was 7.64 and $>1.2 \times 10^{10}$ cells/ml, respectively.

The main objective of the present work was to isolate bacterial cultures capable of degrading SCN⁻from the aqueous industrial wastes. In order to accomplished this objective, activated sludge and garden soil was subjected to the most powerful tool called 'enrichment culture', which is popularly being used world across by the microbiologists to selected desired type of microorganism. There are reports of successfully utilizing this tool for the isolation of microorganism capable of degrading toxic and hazardous chemicals like SCN⁻ and M_xCN (Patil, 1999; Patil, 2008a; Sorokin et al., 2001). Reduction in time during each subsequent transfer could be explained by the fact that bacterial flora in the enrichment medium got gradually acclimatized to the hazardous chemical environment. High bacterial population ($>10^8$ cells/ml) in both the procured samples indicated the presence of substantial organic matter content and nutrient availability, thus giving enhanced probability to obtain SCN⁻degrading cultures.

Purification and Identification of Bacterial Cultures

In all, six heterotrophic bacterial cultures (three each from both enrichment cultures) capable of degrading SCN⁻ were isolated by enrichment technique subsequently followed by streak plate and spread plate technique that were employed for purification of cultures. Microscopic examination showed that all the six bacterial culture were Gram-negative rods and motile. Detailed cultural characteristics were previously reported by Patil (2008a). Based on cultural and biochemical characteristics, all the six identified bacterial cultures belonged to the genus *Pseudomonas* as and reported by Patil (2008b).

The microbial source employed for enrichment culture for the isolation of thiocyante degrading microorganisms were garden soil and sewage sludge of STP. These sites did not have any past history of cyanide or SCN⁻ contamination. The prime objective was to test whether SCN⁻ degrading bacterial cultures could be isolated from such non-contaminated sites and secondly to conduct a comparative assessment of the cultures isolated from two completely different niche areas. The fact that six SCN⁻degrading cultures could be isolated from these samples indicates that SCN⁻ degradation is an intrinsic property of certain microorganisms and that no prior exposure is required to induce this property. SCN⁻degrading ability of various heterotrophic and autotrophic microorganisms have been reported by few authors (Kwon *et al.*, 2002; Patil, 2006; Sorokin *et al.*, 2001; Stratford *et al.*, 1994).

SCN⁻ Degradation Efficiency of the Isolated Bacterial Cultures

Data in Table 1 depicts the wide variation of SCN⁻ degradation efficiency of the bacterial isolates. However, the bacterial consortium isolated from garden soil and activated sludge showed maximum degradation of SCN⁻ (>99.9%) in 42 and 36 h giving the SCN⁻ degradation rate constant (k) of 0.0931 and 0.1086 per h, respectively. In contrast, isolate-2 degraded only 75.9% of SCN⁻ in 48 h (k = 0.029 per h). It was also observed that the first order rate constant of bacterial consortiums was 2-3 folds higher than their individual isolates.

Table 1: SCN⁻ degradation efficiency of pure and mixed bacterial cultures (Conditions: pH 7.0; Temperature 30°C; Inoculum size 10^5 cells/ml; SCN⁻ conc. 50 mg/L; Glucose 10 mM; Agitation speed: 150 rpm; Incubation time: 48 h) (Patil, 2008b)

Source	Bacterial Isolates	% SCN⁻ degradation		Rate of Reaction (mg/l/h)	First Order Rate constant (per h)
		With culture	Control (without culture)		
Bacterial cultures isolated from garden soil	Pseudomonas sp. # 1	78.24	0	0.815	0.0317
	Pseudomonas sp. # 2	75.91	0	0.790	0.0297
	Pseudomonas sp. # 3	82.47	0	0.859	0.0362
	Bacterial consortium (1+2+3)	>99.9 (in 42 h)	0	1.189	0.0931
Bacterial cultures isolated from activated sludge	Pseudomonas sp. # 4	92.07	0	0.959	0.0528
	Pseudomonas sp. # 5	87.65	0	0.913	0.0435
	Pseudomonas sp. # 6	89.41	0	0.931	0.0467
	Bacterial consortium (4+5+6)	>99.9 (in 36 h)	0	1.387	0.1086

This experiment was conducted to ascertain the efficacy of bacterial cultures in their individual capacity and in consortium form. And the results clearly revealed that that consortium of bacteria were efficient compared to individual (pure) isolates (Table 1). These results confirmed the studies carried out by Patil and Picnicker (2000a) on biodegradation of copper- and zinc-cyanide using bacterial consortium. In this study, the consortium consisted of four bacterial isolates out of which three were *Pseudomonas* sp. and one was *Citrobacter* sp. The wide variation in SCN⁻ degradation efficiency among the cultures tested in the present study could be a manifestation of the natural diversity. Bacterial consortium isolated from activated sludge was more efficient than the consortium isolated from garden soil. Inoculated controls did not show any decrease in SCN⁻ levels confirmed that biodegradation of SCN⁻ was the predominant reaction taking place during SCN⁻ degradation by the cultures. Experimental determination of reaction rate and first order rate constants are essentially needed because such data gives valuable information regarding the time requirement for reaction completion and the size of the treatment facilities that must be provided (Patil, 2008b; Sellers, 1999).

Utilization of SCN⁻ as the Sole Source of Cellular Nitrogen by Bacterial Consortium

This experiment was carried out only on bacterial consortium isolated from activated sludge because it was more efficient than the consortium isolated from garden soil. Detailed results of this experiment could be obtained from Patil (2011). Overall results are summarized as follows. It was established that when SCN⁻ was supplemented in M-9 MSM as the sole carbon and nitrogen, the consortium failed to utilize SCN⁻. The SCN⁻ concentration of 50 mg/l remained unchanged throughout the tested period of 48 h. However, it was found that the bacterial consortium was capable of utilizing SCN⁻ as the sole source of cellular nitrogen in the presence external carbon source like glucose (10 mM) within 40 h with an efficiency of >99.9%. This was also confirmed from the control experiments run simultaneously. In the third combination, SCN⁻ when supplemented as the sole carbon source in the presence of external nitrogen like ammonium chloride resulted in complete cessation of consortium growth. In the fourth combination, when SCN⁻ was supplied in MSM along with external carbon (glucose) and nitrogen (ammonium chloride) source, showed an interesting diauxic growth (dioxide) pattern of the bacterial consortium as shown in Fig. 2. The bacterial consortium preferentially utilized ammonium chloride first until its depletion and only then switched over to the utilization of SCN⁻ as nitrogen source.

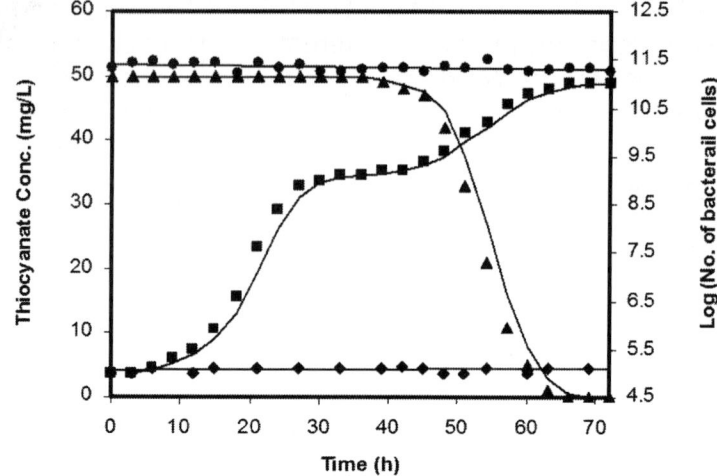

Figure 2: Diauxic growth pattern exhibited by bacterial consortium in the presence two nitrogen sources (viz. SCN⁻ and ammonium chloride) in the presence external carbon source (glucose). Growth of consortium (■) and SCN⁻ degradation (▲) in the presence of two N sources; SCN⁻ concentration in absence of consortium (●); Cessation of bacterial growth in absence of either nitrogen or carbon source (◆) (Patil, 2011).

All bioremediation processes essentially depends on the availability of principal nutrients in the wastes that could potentially be utilized by the microorganisms as either carbon and/or nitrogen source. Elucidating this is crucial because if SCN⁻ is utilized by bacterial consortium as both carbon and nitrogen source, then at practical scale external supplementation of nutrients will not be required, thereby benefiting the industries economically those using microbial technologies for the effluents containing SCN⁻, cyanide and metal-cyanides. However, in the present study, the SCN⁻ compound posed toxic problems to the consortium for growth utilizing it as suitable growth substrate. It is well known that concentration of nitrogen required for a given amount of growth is less than the requirement for carbon it might be easier for bacterial consortium to utilize SCN⁻ as the source of nitrogen in the presence of a separate source of carbon and energy (Patil, 1999). Therefore, enrichment culture was designed/manipulated for the isolation of microorganisms capable of degrading SCN⁻ as the source of nitrogen nitrogen (Patil, 2006; Patil, 2008a). The experiments conducted explicitly proved that SCN⁻ is used by the consortium as nitrogen source in the presence of external carbon viz. glucose, thereby giving the C/N ratio of 10. In view of microbial process development, it is imperative to supplement some cheaper source of carbon like molasses, which is readily available in India at cheaper rate. Patil (1999) had successfully demonstrated the use of molasses as carbon source to develop a microbial technology for metal cyanide biodegradation/removal from wastes utilising it as the sole nitrogen source. There are few reports, which describe microbial SCN⁻ degradation utilising it as the sole nitrogen source (Bipinraj *et al.* 2003; Patil, 2011; Sorokin *et al.* 2001). The bacterial consortium ceased to grow when SCN⁻ was supplied as sole carbon source in the presence of external nitrogen. This could be attributed to the higher amount of available nitrogen compared to carbon (C/N ratio 0.5). Obviously the culture would find it more difficult to obtain sufficient amount of energy from low amount of carbon.

Example of diauxie pattern (biphasic growth) in *Escherichia coli* in the presence of two carbon sources (viz. glucose and lactose) is well documented (Atlas, 1997). In the present study, diauxic growth pattern was observed when two nitrogen salts (i.e. SCN⁻ and ammonium chloride) along with one carbon source (glucose) were supplied to the consortium. Ammonium chloride acted as preferred growth substrate by the consortium followed by SCN⁻ degradation. This suggests that SCN⁻ utilization by consortium culture is inducible. It was also revealed that biodegradation of SCN⁻ took place rapidly (within 25 h) in second phase of growth after the exhaustion of ammonium chloride from medium in the first phase. The biomass that built-up in the first phase of growth was readily available in the medium for SCN⁻ degradation in the second phase, which ultimately led to the rapid biodegradation of SCN⁻. This result immediately suggests its possible application in bioreactor designing that will retain large microbial biomass. Immobilization of the biomass in bioreactor using inert material will certainly hasten the process of biodegradation of toxic SCN⁻.

Further, it could be also observed from the experiments that decrease of SCN⁻ concentration in the MSM was concomitant with the increase in bacterial population. The fact that the final cell density obtained was considerably high (>10⁸cells/ml) indicated the use of well acclimatized SCN⁻ tolerant culture, having high SCN⁻ removal efficiency and therefore has immense potential of using the microbial technology on industrial scale. The treatment of wastewater involves a number of chemical and biological reactions and conversions. The rate at which these reactions and conversions occur decides the size of the treatment facilities that must be provided (Tchobanoglous and Burton, 1997). The study also showed that SCN⁻ degradation by the bacterial consortium isolated from activated sludge was comparatively more efficient than the consortium isolated from garden soil. This might be due to the acclimation/tolerance of sewage microorganisms to a variety of hazardous and non-hazardous waste contaminants/components naturally existing in it, thereby making them more tolerant and efficient degraders as compared to the microbial flora prevailing in garden soil.

Factors Influencing SCN⁻ Biodegradation

Factors influencing SCN⁻ biodegradation was restricted to the bacterial consortium isolated from activated sludge because of its high degradation efficiency compared to the consortium isolated from garden soil as mentioned earlier. Table 2 shows that degradation of SCN⁻ was significantly influenced by the various factors tested. Optimum pH and temperature for maximum SCN⁻ biodegradation (>99.9%) was found to be 7.0 and 30°C, respectively. Under the optimized conditions of pH and temperature, the initial cell density had a substantial influence on the biodegradation efficiency of SCN⁻. With initial cell density of 10^8 cells/ml, SCN⁻ degradation process completed within 24 h with >99.9% efficiency. As regard to carbon source, the consortium culture exhibited maximum biodegradation efficiency only above glucose concentration of 5 mM.

Table 2: Degradation of SCN⁻ by a bacterial consortium isolated from activated sludge as a function of pH, temperature, initial cell density and glucose

Parameter	Range selected	% SCN-biodegradation	Parameter	Range selected	% SCN-biodegradation
pH	Control*	0	Cell Density (cells/ml)	Control	0
	5	35.3		10^5	19.5
	5.5	43.0		10^6	47.1
	6	75.3		10^7	62.8
	6.5	84.6		10^8	>99.9
	7	>99.9		10^9	>99.9
	7.5	89.2	Glucose (mM)	Control	0
	8	87.6		0.1	6.2
	8.5	84.5		1	24.7
	9	84.2		5	>99.9
	9.5	56.9		10	>99.9
Temperature	Control	0		20	>99.9
	10	16.3			
	20	51.8			
	30	>99.9			
	37	36.3			
	45	0			

[i] - * Control indicates flask without culture; All the values are the average of two readings

In general, the SCN⁻ containing effluents released from various industries have pH in neutral to alkaline range. Our study showed that growth of SCN⁻ degrading bacterial consortium occurred in wide range of pH (6.0-9.0), while the optimum being 7.0. From practical applicability point of view very little or no pH adjustment would be required for the effluents containing SCN⁻. Experiments also showed the unchanged pH of the solution after SCN⁻ biodegradation. This may be perhaps due to the formation of ammonia as one of the by-products of SCN⁻ degradation, which neutralized the accumulated carboxylic acids in the medium. These results corroborate with the studies carried out by other researchers on the biodegradation/bio detoxification free cyanide (Babu et al., 1993) and metal-cyanides (Patil and Paknikar, 2000b). The bacterial consortium used in the study was neutrophil and mesophilic. Optimum temperature for thiocyanate degradation by bacterial consortium was 30°C. This is very important from the view point of actual applicability of the bioremediation process in a tropical country like India having ambient temperature ranging from 20-40°C. Results on the impact of inoculum size clearly showed that SCN⁻ degradation increased with the increase in inoculum size. Therefore, from the point of view of process development, it is essential to use a reactor capable of retaining high microbial biomass that will hasten the degradation of SCN⁻. Results on the influence of glucose requirement for SCN⁻degradation could possibly be explained on the basis of nutrient availability. Even though the available nitrogen in the form of SCN⁻ was ample the externally supplied glucose at concentrations < 5 mM limited the biodegradation process. However, at adequate glucose concentration of ≥ 5 mM complete degradation could take place. These findings confirm the utilization of SCN⁻ as the sole source of nitrogen. In the previous studies carried out by Stratford et al. and Wood et al. glucose was supplemented at the concentration of 10 and 25 mM for the degradation of 3 mM (≈ 174 mg/l) and 2.5 mM (≈ 145 mg/l) of SCN⁻, respectively (Stratford et al., 1994; Wood et al., 1998). However, these reports did not mention optimization of this parameter, which needs to be worked out for economizing the process. In another study carried out by Patil (1999) on the biodegradation of various metal-cyanides (copper-, nickel-, zinc- and silver-cyanide), glucose was required in the range of 1-5 mM (≈ COD 100 – 500 mg/l) (Patil, 1999). Scanty information is available on the biochemical pathway involved in SCN⁻ biodegradation. For heterotrophic bacterium, Stratford et al.

has proposed the conversion of SCN⁻ to carbon dioxide and ammonia via cyanate by an inducible enzyme; while the sulphur moiety gets hydrolyzed to sulphide, which further gets oxidised to tetrathionates via formation of thiosulphate (Stratford *et al.*, 1994). It might be possible that bacterial cultures isolated in the present study also have similar SCN⁻ removal or tolerant mechanism.

Biosorption of SCN⁻ by Bacterial Consortium at High Cell Density

It can be seen from Table 3 that the bacterial consortium had low bio sorption efficiency (~7-14%) at the pH values tested (6.5 to 7.5). In fact it is possible that bios orbed SCN⁻ also could subsequently be biodegraded by the live culture used in the biodegradation process. These observations confirmed that biodegradation of SCN⁻ was the predominant reaction taking place during detoxification of SCN⁻ by the consortium culture isolated from activated sludge.

Table 3: Biosorption of SCN⁻ by consortium culture

pH	SCN⁻ concentration (mg/l)		% Sorption
	Initial	Final	
6.5	53.29	48.22	9.51
7.0 (optimum pH)	50.94	45.97	13.68
7.5	51.03	47.50	6.91

[i] - *Optimum pH

Impact of Cations and Anions on SCN⁻ Biodegradation

Apart from SCN⁻ various metal cations and anions are normally present in the various industrial effluents. Therefore, the influence of some of the commonly occurring cations such as copper, cadmium, iron, lead, nickel, zinc and anions such as sulfates, chlorides and cyanide on SCN⁻ biodegradation was checked.

Table 4 shows the effect of various cations such as copper, nickel, zinc, cadmium, iron and lead on biodegradation of SCN⁻ in thiocyanate-metal system. It can be seen that biodegradation of thiocyanate was not affected in the presence of copper, nickel and zinc (degradation >90%). In presence of lead and cadmium the biodegradation efficiency was reduced by approximately 30 and 45%, respectively. Chromium and iron significantly affected the degradation of SCN⁻ by >80%. Anions such as sulfates and chlorides (1000 µM and 10000 µM, respectively), and cyanide (0.1-1 mM) however, did not had much impact on SCN⁻ degradation.

Table 4: Effect of cations and anions onµSCN⁻ biodegradation

Thiocyanate + cations/anions	% Thiocyanate biodegradation
Thiocyanate (Control without culture)	0
Thiocyanate (Control with culture)	>99
Thiocyanate + Copper	95.2
Thiocyanate + Nickel	>99
Thiocyanate + Zinc	90.4
Thiocyanate + Cadmium	56.2
Thiocyanate + Iron	17.9
Thiocyanate + Lead	69.5
Thiocyanate + Chromate	11.8
Thiocyanate + Sulfate	74.0
Thiocyanate + Chlorides	90
Thiocyanate + Cyanide	>98

Biodegradation of the SCN⁻ was adversely affected in the presence of metals such as iron and chromium. In case of free cyanide, it is known that cyanide ion has a great tendency to act as a ligand and can thus be found associated with metal-complexes (Pohlandt et al., 1983). Cyanide complexes with different metals have widely varying stabilities depending on metal oxidation states (Cotton and Wilkinson, 1972), but can be broadly classified as weakly complexes cyanides and strongly complexes cyanides. The former group includes complexes with copper, cadmium, lead, nickel and zinc (Chapman, 1992) while the latter group consists of very stable hexacyanoferrates and chromium-cyanide (Lordi et al., 1980). Since the chemistry of free cyanide and SCN⁻ almost being similar, the heavy metals are capable of forming ligands with thiocyanate to form metal-thiocyanate complexes. The high stability of iron-thiocyanate and chromium- SCN⁻ might be the reason for poor degradation efficiency observed in presence of these ions.

Degradation of SCN⁻ from Industrial Effluent by Bacterial Consortium

The SCN⁻ containing effluent simulated in laboratory could be effectively treated by the bacterial consortium with a degradation efficiency exceeding >99.9%. The time incurred for the complete biodegradation of SCN⁻ from waste waters was less than 24 h. Table 5 shows the parameters such as pH, total cyanide, COD and metal content before and after biodegradation. COD removal was more than 80%. It was also observed that the level of soluble copper, zinc, silver and nickel was reduced to less than 5 mg/l. There was no significant change in the pH after biodegradation.

Table 5: Composition and biodegradation of simulated industrial effluent

| Parameter | Simulated Industrial Effluent concentration (mg/l) | | | % Removal efficiency |
	Before biodegradation	After biodegradation	Uninoculated control	
Color	Colourless	Colourless	Colourless	-
Turbidity	Clear	clear	Clear	-
pH	7.3	7.7	7.4	-
Thiocyanate	51.5	< 0.1	52.4	>99
Copper	12.5	1.92	13.1	84.6
Nickel	8.1	0.55	8.0	93.2
Zinc	18.3	2.17	16.9	88.1
Iron	3.9	0.1	4.1	97.4
Sulfates	55	57	62	-
Chlorides	42	45	50	-
Cyanide	5.2	0.13	4.7	97.5
COD	< 550	97	600	82.3

[i] - All the values given in the table are in mg/l, except pH

The microbial process for degradation of thiocyanate was found to be highly effective in the detoxification of simulated industrial effluents. The levels of total thiocyanate, COD and metals could be brought down below the statutory limits as per Indian Standards (IS: 2490-1981). The treatment of effluent required more time as compared to the synthetic solutions. It is known that the applicability of any such process to real effluents is always complicated by the fact that effluents contain a variety of other contaminants which might interfere with or prolong the detoxification process. However, it must be emphasized that the microbial process described was highly efficient, safe and environment-friendly. In addition, the process had the potential of becoming economically attractive if scaled-up to a sufficient level, especially as a continuous operation. Therefore, it was decided to further develop the process in continuous mode and evaluate its performance.

Treatment of SCN⁻ Waste in a Continuous Treatment System (CTS)

Studies in CTS showed that SCN⁻ level in the treated effluent was consistently below 0.1 mg/l for over a period of 30 days. The HRT of CTS was constant during the treatment period around 20 h. A closer monitoring of the CTS revealed that further reduction in the HRT was not possible because the bacterial cells could not be retained in the system. The COD removal efficiency after treatment was >75% for thiocyanate effluent (Table 6).

Table 6: Treatment of metal-cyanide waste waters in CTS

Parameter	Influent	Effluent	Bureau of Indian Standards (BIS)
pH	7.0 - 7.3	7.5 - 7.7	5.5 - 9.0
Thiocyanate	51.67 ± 2.1	0.03± 0.01	NA
COD	596 ± 103	147 ± 41	250

[i] - All the figures in the table are expressed in mg/l, except pH; *HRT of the system was ~20 h. Figures represent average values of 30 readings taken each at 24 h interval

The results of CTS showed that SCN⁻ was degraded efficiently by the bacterial consortium with the minimum hydraulic retention time (HRT) of approximately 20 h. However, there was no reduction in the HRT of CTS further. The main reason for this was the continuous loss of active biomass from the reactor, which makes it unattractive from process economics point of view. This necessitates the immobilization of the bacterial consortium in the reactor. In principle, it is possible to retain active biomass in CTS if the culture used has an inherent property of producing wall growth. Further, biomass retention is also possible by changing the reactor design, introducing inert support material or changing nutrient supplementation, etc. However, the consortium culture used in the present studies did not show wall growth. Also, in our studies during optimization of process parameters it was conclusively proved that degradation efficiency increased with the increase in cell number, which in turn hastens the degradation of SCN⁻. Thus, the above results emphasize the fact that the bioremediation process developed during the course of present work is highly efficient and completely safe. After further scale-up the bacterial process developed could have the following advantages: (i) no sludge generation; (ii) no expensive chemical additives required; (iii) very little or no pH adjustment required; (iv) the process would be easy to operate and maintain. Thus, the bacterial process developed could have the potential of becoming an economical and reliable alternative to the conventional processes employed for the treatment of SCN⁻ bearing industrial effluents on a commercial scale.

ACKNOWLEDGEMENTS

The author gratefully acknowledges the research grant provided by University Grants Commission (UGC), WRO, and Pune.

REFERENCES

1.   APHA, AWWA, WEF (1998) *"Standard Methods for the Examination of Water and Wastewater"*, 20th Ed. American Public Health Association, Washington, DC.

2.   Atlas, R.M. (1997) *Principles of Microbiology*, McGraw-Hill, New York.

3.   Babu, G.R.V., Wolfram, J.H. and Chapatwala, K.D. (1993) Degradation of Inorganic Cyanides by Immobilised *Pseudomonas putida* Cells, In: Torma, A.E., Apel, M.L. and Brierley, C.L. eds., Biohydrometallurgical Technologies, The Minerals Metals and Materials Society, Warrendale, PA, Vol. II, pp. 159-165

4.   Bauchop, T. and Elsden, S.R. (1960) the growth of microorganisms in relation to energy supply, Gen. Microbiol. 23: 457.

5.   Beekhuis, H.A. (1975) Technology and Industrial Application In: Newmann, A.A., Chemistry and Biochemistry of Thiocyanic Acid and its Derivatives, Academic Press, London.

6.   Bipinraj, N.K., Joshi, N.R. and Paknikar, K.M. (2003) *"Biohydrometallurgy: A Sustainable Technology in Evolution"*, In: International Biohydrometallurgy Symposium, Athens, pp. 491.

7.   Chaudhari, A.U. and Kodam, K.M. (2010) Biodegradation of thiocyanate using co-culture of *Klebsiella pneumoniae* and *Ralstonia* sp. *Applied Microbiology and Biotechnology* 85(4): 1167-1174.

8.   Cotton, F.A. and Wilkinson, G. (1972) *Advanced Inorganic Chemistry: A Comprehensive Text*. Wiley-Interscience, New York.

9.   Dash, R.R., Gaur, A. and Balomajumder, C. (2009) Cyanide in industrial wastewaters and its removal: A review on biotreatment. *Journal of Hazardous Materials*, 163(1):1-11.

10.   Gaddi Shivanand S and Patil Yogesh B (2011) Screening of some low-cost waste biomaterials for the sorption of silver-cyanide [Ag (CN) 2-] from aqueous solutions. *International Journal of Chemical Sciences* 9: 1063-1072.

11.   Gurbuz, Fatma, Ciftci, Hasan, Akcil, Ata and Karahan, Aynur Gul. (2004) Microbial detoxification of cyanide solutions: a new biotechnological approach using algae, *Hydrometallurgy*, 72, 167-176.

12.   Holt, J.G. (1989) *Bergey's Manual of Systematic Bacteriology*. Vol. I. Williams and Wilkins, Baltimore, USA.

13.   Hughes, M.N. (1975) General Chemistry. In: Newmann, A.A., ed., Chemistry and Biochemistry of Thiocyanic Acid and its Derivatives, Academic Press, London.

14. Hung, C.H. and Pavlostathis, S.G. (1998) Fate and transformation of thiocyante and cyanate under methanogenic conditions. *Applied Microbiology and Biotechnology* 49:112-116.

15. Karavaiko, G.I., Kondrat'eva, T.F., Savari, E.E., Grigor'eva, N.V. and Avakyan, Z.A. (2000) Microbial degradation of cyanide and thiocyanate,*Microbiology*, 69(2):167-173.

16. Kelly, D.P. and Baker S.C. (1990) Theorganosulphur cycle: aerobic and anaerobic processes leading to turnover of C1-sulfur compounds. FEMS Microbiology Review 87: 241-246.

17. Kwon, Hyouk K., Woo, Seung H. and Park, Jong M. (2002) Thiocyanate degradation by *Acremonium strictum* and inhibition by secondary toxicants,*Biotechnol. Lett.*24, 1347-1351.

18. Lanza, M.R.V. and Bertazzoli, R. (2002) Cyanide Oxidation from Wastewater in a Flow Electrochemical Reactor. *Industrial Engineering and Chemical Research*. 41, 22-26.

19. Lordi, D.T., Lue-Hing, C., Whitebloom, S.W., Kelada, N. and Dennison, S. (1980) Cyanide problems in municipal wastewater treatment plants. *J. Wat. Pollut. Cont. Fed.* 52: 597-609.

20. Millar, J.H. (1972) *"Experiments in Molecular Genetics"*, Cold Spring Harbour, NY: Cold Spring Harbour Laboratory.

21. Mudder, T.I. and Whitlock, J.L. (1984) Biological treatment of cyanidation waste waters. *Minerals and Metallurgical Processing*, pp. 161-165

22. Patil Y.B. (1999) Studies on Biological Detoxification of Metal-Cyanide Containing Industrial Effluents, Ph.D. Thesis, University of Pune, Pune, India.

23. Patil YB and Paknikar KM (1999a) Removal and recovery of metal cyanides using a combination of biosorption and biodegradation processes.*Biotechnology Letters* 21: 913-919.

24. Patil YB and Paknikar KM (1999b) Removal and recovery of metal cyanides from industrial effluents. *Process Metallurgy* 9: 707-716.

25. Patil YB and Paknikar KM (2000a) Development of a process for biodetoxification of metal cyanides from wastewaters. *Process Biochemistry* 35: 1139-1151.

26. Patil YB and Paknikar KM (2000b) Biodetoxification of silver-cyanide from electroplating industry wastewater. *Letters in Applied Microbiology* 30: 33-37.

27. Patil YB and Paknikar KM (2001) Biological detoxification of nickel-cyanide from industrial effluents. *Process Metallurgy* 11: 391-400.

28. Patil Yogesh B (2006) Isolation of thiocyanate degrading chemoheterotrophic bacterial consortium. *Nature Environment and Pollution Technology*5 (1): 135-138.

29. Patil Yogesh B (2008a) Biodegradation of thiocyanate from aqueous waste by a mixed bacterial community. *Research Journal of Chemistry and Environment* 12(1): 69-75.

30. Patil Yogesh B (2008b) Thiocyanate degradation by pure and mixed bacterial cultures. *Bioinfolet* 5(3): 308-309.

31. Patil Yogesh B (2011) Utilization of thiocyanate (SCN-) by a metabolically active bacterial consortium as the sole source of nitrogen. *International Journal of Chemical, Environmental & Pharmaceutical Research*, 2: 44-48.

32. Patil Yogesh B (2012) Development of an innovative low-cost industrial waste treatment technology for resource conservation - A case study with gold-cyanide emanated from SMEs. *Procedia-Social and Behavioral Sciences* 37: 379-388.

33. Patil Yogesh B and Kulkarni Anil R (2008) Environmental sensitivity and management of toxic chemical waste in mining industry with special reference to cyanide. In: *High Performing Organizations: Needs and Challenges*, Tata McGraw Hill Publications, Part D: pp. 183-196.

34. Patil Yogesh, Chourey Jayati and Rao Prakash (2012) Biotechnological strategy for the management of industrial waste with concurrent mitigation of global warming – A feasibility study using microalgae. In: *Inclusiveness and Innovation – Challenges for Sustainable Growth of Emerging Economies* (Edited by Rajiv Divekar and BR Londhe), Excel India Publishers, New Delhi, India.

35. Prashanth MS and Patil Yogesh B (2006) Behavioural surveillance of Indian major carp *Catlacatla*(Hamilton) exposed to free cyanide. *Journal of Current Sciences* 9(1):313-318.

36. Sellers, K. (1999) *"Fundamentals of Hazardous Waste Site Remediation"*, Lewis Publishers, Washington, DC.

37. Sorokin Dimitry Y., Tourova, Tatyana P., Lysenko, Anatoly M. and Kuenan, J. Gigs. (2001) Microbial thiocyanate utilization under highly alkaline conditions.*Applied and Environmental Microbiology* 67(2): 528-538.

38. Stratford, J., Dias, A.E.X.O and Knowles, C.J. (1994) the utilization of thiocyanate as a nitrogen source by a heterotrophic bacterium: the degradative pathway involves formation of ammonia and tetrathionate, *Microbiology* 140: 2657-2662.

39. Tchobanoglous, G. and Burton, F.L., (1997) Wastewater Engineering: Treatment, Disposal, Reuse, Metcalf and Eddy Inc., McGraw-Hill Publication, New Delhi.

40. Thakur Ravindra Y and Patil Yogesh B (2009) Management of thiocyanate pollution using a novel low cost natural waste biomass.*South Asian Journal of Management Research* 1(2): 85-96.

41. Van Zyl, Andries W., Harrison, Susan T.L., van Hille, Robert P. (2011) Biodegradation of thiocyanate by a mixed microbial population. In: Mine Water – Managing the Challenges (IMWA 2011), Aachen, Germany.

42. Westley, J. (1981) *Cyanide and sulfane sulfur*. In: Vennesland, B., ed., Cyanide in Biology, Academic Press, London, pp. 201-212.

43. Wood, A.P., Kelly, D.P., McDonald, I.R., Jordan, S.L., Morgan, T.D., Khan, S., Murell, J.C. and Borodina, E., (1998) A novel pink-pigmented facultative methylotroph, *Methylobacterium thiocyanatum* sp. nov., capable of growth on thiocyanate as sole nitrogen source. *Arch. Microbiol.*, 169, 148-158.

44. Wood, J.L. (1975) Biochemistry, In: Newmann, A.A., ed., Thiocyanic Acid and its Derivatives, Academic Press, London, United Kingdom.

45. Zargury, G.J., Oudjehani, K. and Deschenes, L. (2004) Characterisation and availability of cyanide in solid mine tailings from gold extraction plants.*Science of the Total Environment* 320: 211-224.

Adsorption of Hg(II) from Aqueous Solution Using Adulsa (Justicia adhatoda) Leaves Powder: Kinetic and Equilibrium Studies

Mohd Aslam[1,2], Sumbul Rais[2], Masood Alam[2], and Arulazhagan Pugazhendi[1]

[1]Center of Excellence in Environmental Studies, King Abdulaziz University, Jeddah 21589, Saudi Arabia

[2]Department of Applied Sciences and Humanities, Faculty of Engineering and Technology, Jamia Millia Islamia, New Delhi-110025, India

ABSTRACT

The ability of Adulsa leaves powder (ALP) to adsorb Hg (II) from aqueous solutions has been investigated through batch experiments. The ALP biomass was characterized by Fourier transforms infrared spectroscopy and scanning electron microscopy. The experimental parameters that were investigated in this study included pH, adsorbent dosage, and effect of contact time along with initial metal ion concentration. The adsorption process was relatively fast, and equilibrium was achieved after 40 min of contact time. The maximum removal of Hg(II), 97.5% was observed at pH 6. The adsorption data were correlated with Langmuir, Freundlich, and Temkin isotherms. Isotherms results were amply fitted by the Langmuir model determining a monolayer maximum adsorption capacity (q_m) of ALP biomass equal to 107.5 mg g^{-1} and suggesting a functional group-limited sorption process. The kinetic process of Hg(II) adsorption onto ALP biomass was tested by applying pseudofirst-order, pseudosecond-order, Elovich, and intraparticle-diffusion models to correlate the experimental data and to determine the kinetic parameters. It was found that the pseudosecond order kinetic model for Hg(II) adsorption fitted very well. The rate determining step is described by intraparticle diffusion model. These studies considered the possibility of using Adulsa plant leaves biomass as an inexpensive, efficient, and environmentally safe adsorbent for the treatment of Hg(II) contaminated wastewaters.

INTRODUCTION

The presence of toxic heavy metals in aqueous stream, arising from the discharge of untreated metal containing effluents into water bodies has become one of the most important environmental issues in the past few decades. Mercury is an element, and it cannot be created by people, nor can it be destroyed. Mercury is released into the environment by volcanic eruptions, and it naturally occurs in the earth's crust, often in the form of mercury salts such as mercury sulfide [1].

Mercury, amongst other heavy metals has attracted global concern due to its extensive use, toxicity, wide spread distribution, and the biomagnifications. Several kinds of human activities (anthropogenic) release mercury into the environment that include effluents from paint and chloralkali, pulp paper, oil refining, battery production, fossil fuel burning, mining and metallurgical processes, rubber processing, and fertilizer industries [2–5]. Other major source of mercury emission into the atmosphere is flue gases from coal combustors used in electricity generation [6, 7]. The European Union considers mercury as a priority and hazardous pollutant and defines a maximum permissible concentration of total mercury as low as 1 μgL^{-1} for drinking water and 5 μgL^{-1} for wastewater discharge [8]. The primary targets for toxicity of mercury and mercury compounds are the nervous system, the kidneys, and the cardiovascular system. It is generally accepted that developing organ systems (such as the fetal nervous system) are the most sensitive to toxic effects of mercury. Other systems that may be affected include the respiratory, gastrointestinal, hematologic, immune, and reproductive systems [9].

Conventional treatments to remove mercury from aqueous solution include precipitation, electrolysis, ion exchange, adsorption, cementation, liquid membranes, and liquid-liquid extraction [10, 11]. However, these processes are ineffective at low metal concentration or expensive due to toxic sludge disposal, chemical reagents for metal recovery, and sorbent regeneration and high energy requirements. Therefore, more effective low cost alternatives are urgently required. In the past two decades, biosorption received a considerable attention for the removal of heavy metals from waste water [12–17].

Various potentially inexpensive adsorbents such as bark [18], Carica papaya [19], Sewage sludge [20], wheat bran [21], walnut shell [22], coffee grounds [23], waste rubber [24], coconut husk [25], fertilizer waste slurry [26], algae [27], peanut hull [28], jackfruit peel [29], coal-fly ash [30], coir pith [31], and sago waste [32] have been used for the removal of mercury from aqueous solution.

Adulsa (Justicia adhatoda) is a small evergreen herbal plant in the family Acanthaceae. It is distributed all over the plains of India and in lower Himalayan ranges. The leaves of the plant contain an essential oil and alkaloids vasicine, N-oxides of vasicine, vasicinone, deoxyvasicine, and maiontone. The plant has been recommended by Ayurvedic physicians for the management of various types of respiratory disorders. It possesses potent bronchodilatory, expectorant, antispasmodic and antiseptic properties. The aim of the present study is to evaluate the potential of Adulsa leaves powder (ALP) biomass for the adsorption of mercury from aqueous solution by batch operation technique. The effects of optimum biosorption conditions such as pH, initial metal ion concentration, contact time, and biomass dosage have been explored. The Langmuir, Freundlich, and Temkin models were used to describe the equilibrium isotherms. To correlate the experimental data and to determine the kinetic parameters, pseudofirst order, pseudosecond-order, intraparticle diffusion, and Elovich model were evaluated.

EXPERIMENTAL

Adsorbate Preparation

The stock aqueous solutions of desired concentration have been prepared by dissolving the appropriate amount of $Hg(NO_3)_2.H_2O$ in double distilled water (DDW). The stock solution was used to prepare dilute solutions of different working concentrations. All the chemicals used in this study were of analytical grade from Merck Company (Darmstadt, Germany).

Adsorbent Preparation

Adulsa leaves powder (ALP) used as an adsorbent were collected from Horticulture Department, Jamia Millia Islamia, Central University, New Delhi, India. The plants were washed thoroughly using tap water in order to remove the water soluble impurities and other surface-adhered particles. Only the leaves of the plants were utilized in this study. The washed plant leaves were oven-dried at 80°C, ground using a blender and sieved through a 40–50 mesh BSS screens in order to obtain uniform particle size. Leaves biomass was washed four times with DDW in order to remove soluble material or biomolecules that might interact with any adsorbed metal ions. So obtained biomass was dried in oven at 80°C and stored in desiccators.

Adsorbate Analysis

The final solutions of mercury concentrations of the samples were determined by using a flow injection analysis system, atomic absorption spectrophotometer (FIAS-AAS, Perkin-Elmer model 3100). The analytical wavelength used was 253.7 nm with a slit width of 0.7 nm having hollow cathode lamp current of 6 mA current. Standards were prepared by diluting a 1000 mg/L Hg(II) stock solution with 3%

HCl solution and 2-3 drops of $KMNO_4$, and linear calibration curves were obtained with correlation coefficients of $R^2 = 0.99$ or better. Three replicates of each sample were analyzed, and the mean value was reported.

Batch Adsorption Studies

The batch adsorption studies were carried out in 250 mL Erlenmeyer flasks containing 1 g of the adsorbent in 100 mL of Hg(II) solution at $30\pm2°C$ on a rotary shaker at 150 rpm. The effect of pH on biosorption rate was investigated in a pH range of 3 to 9, which was regulated by microadditions of 0.1 N HCl or 0.1 N NaOH at the beginning of the experiment. The best amount of biomass was determined by changing the biomass dosage from 0.25 g to 1 g in 100 mL of Hg(II) solution. The initial concentration of Hg(II) solution taken for this study was 25, 50, 75, and 100 mg/L. For optimization of contact time, samples were taken at predetermined time intervals (0–120 min) for determination of the residual metal ion concentration in the solution. Before analysis, the samples were centrifuged at 5000 rpm to separate the biomass. The residual metal concentrations in the supernatant were analyzed by FIAS-AAS.

The amount of the metal adsorbed (mg) per unit mass of biomatrix was obtained by using the equation:

$$q_e = (C_o - C_e) \times \frac{V}{m},$$

(1)

where q_e is amount of metal ion adsorbed per gram of biomass ($mg \cdot g^{-1}$), C_o is the initial metal ion concentration ($mg\ L^{-1}$), C_e is the final metal ion concentration ($mg\ L^{-1}$), V is the volume of the reaction mixture in liter, and m is the weight of biomass in the reaction mixture in g.

RESULTS AND DISCUSSION

Characterization of Adsorbent

Fourier transform infrared (FTIR) spectroscopy was done to identify the chemical functional groups present on native ALP and the Hg(II)-loaded ALP. The spectrum was collected by PU420, JASCO spectrometer in the range 400–4,000 cm⁻¹ using a KBr window. The background obtained from the scan of pure KBr was automatically subtracted from the sample spectra. Spectra were plotted using the same scale on the absorbance axis. Various functional groups such as amine (–NH), carboxylate anions (–COO⁻), hydroxyl (–OH), and others: (N=O) (–C=C), (–C–C), (–C=O), (–C–O), (–C–N), and (–C–H) have been proposed to be responsible for the adsorption heavy metal ions on the cell surfaces of adsorbent. Their importance for metal uptake depends on factors such as the quantity of sites, its accessibility and chemical state, or affinity between site and metal. The FTIR absorption spectra of unloaded and Hg(II) loaded were taken (Figures 1(a) and 1(b)) to confirm the presence of different functional groups in adsorbent.

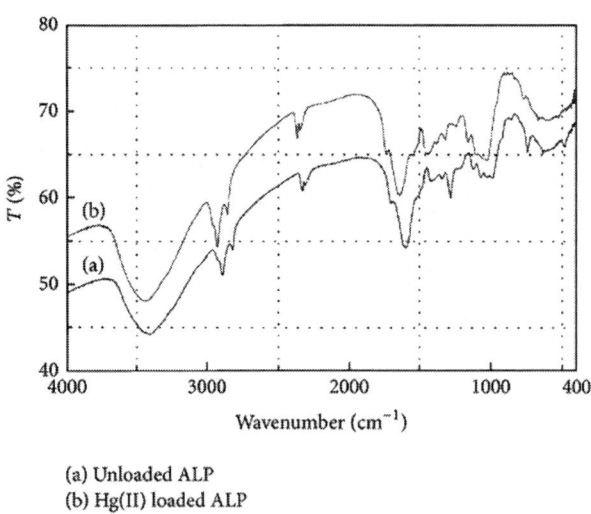

(a) Unloaded ALP
(b) Hg(II) loaded ALP

Figure 1: FTIR spectra's of (a) unloaded ALP and (b) Hg (II) loaded ALP.

In the FTIR absorption, spectra of unloaded ALP biomass show a broadband at 3438 cm^{-1} which indicates the presence of hydrogen-bonded –OH stretching modes from alcohol and phenols and also dominated by –NH stretching. The bands at 2900 and 2853 cm^{-1} in IR spectra of ALP may be due to the C–H stretching vibrations. The peak at 2329 represents stretching vibrations of $-NH_2^+$, $-NH^+$, and –NH groups of the unloaded ALP biomass. The bands appearing at 1629 and 1380 cm^{-1} are attributed to the formation of oxygen functional groups like a highly conjugated C=O stretching in carboxylic groups and N=O bending in nitro groups, respectively. The peak appeared at 1024 cm^{-1} has been assigned to C–O stretching in ethers. The peak at 607 cm^{-1} is caused by C–N–C scissoring, which is found in polypeptide structure.

The small shift was obtained in the absorbance peak of loaded Hg(II) ALP biomass compared with that of unloaded ALP biomass which is shown in Figures 1(a) and 1(b). The broadband observed at 3438 cm^{-1} for hydrogen-bonded –OH stretching and –NH stretching was shifted to 3449 cm^{-1}. The peaks at 2900 and 2853 cm^{-1} due to the C–H stretching vibrations were shifted to 2924 and 2860 cm^{-1}, respectively. The stretching vibration band observed at 2329 cm^{-1} was altered to 2352 cm^{-1}. The peaks of highly conjugated C=O stretching and N=O bending observed at 1629 and 1380 cm^{-1} were shifted to 1646 and 1400 cm^{-1}. C–O stretching peak at 1024 cm^{-1} was changed to 1026 cm^{-1}. The C–N–C scissoring peak at 607 cm^{-1} was also shifted to 780 cm^{-1}. It should also be noted that FTIR results did not provide any quantitative analysis as well as the information about the level of affinity to metal of the functional groups presented in the adsorbents. They only presented the possibility of the coupling between the metal species and the functional group of the adsorbents.

The surface morphology of the ALP and Hg(II)-loaded ALP was analyzed by scanning electron microscopy (SEM) by using JEOL-JSM-6380 model which is shown in Figures 2(a) and 2(b). SEM micrograph of fresh ALP (Figure 2(a)) revealing the nature of biomass which is rough and heterogeneous with considerable amount of voids and lot of ups and downs. The uptake of Hg(II) by ALP is demonstrated by the change in morphology of the adsorbent's surface with the formation of-ike structure (Figure 2(b)). Based on the surface morphology results of ALP, it is suggested that produced ALP can be used as adsorbent for liquid-solid adsorption processes, due to the importance of fibrous material to many liquid-solid adsorption processes.

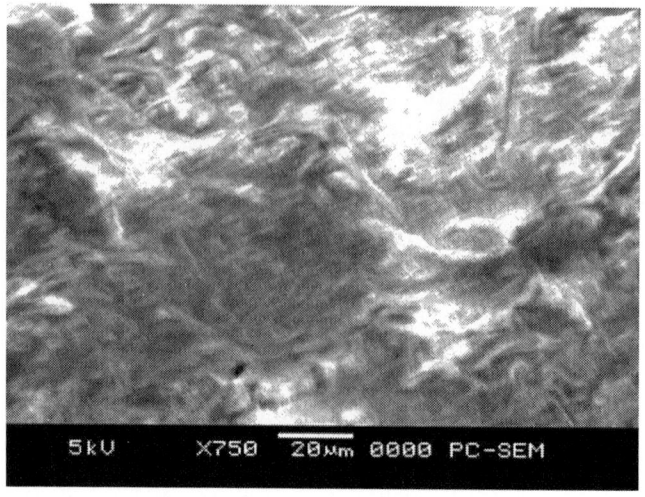

(a)

(b)

Figure 2: SEM of (a) unloaded ALP and (b) Hg (II) loaded ALP.

Effect of pH

The effect of pH was found to be the most important variable governing the biosorption of metal ions by sorbent in the pH range of 3–9. Change in the removal of Hg(II) by ALP biomass with pH is shown in Figure3. A significant increase in Hg(II) uptake was obtained as the pH increases from 3 to 6. The highest adsorption efficiency (97.5%) was observed at pH 6. At low pH value, more proton will be available to protonate the active sites and then less attraction towards the Hg(II) ions (low metal uptake) due to high electrostatic repulsion. When the pH was increased, the competing effect of H^+ ion decreased and the positively charged Hg(II) ion took up the free active sites [45]. At pH greater than 6, the decrease in mercury (II) uptake occurs due to the formation of hydroxyl species such as [Hg(OH) or $Hg(OH)_2$], and competing between mercury ions and hydroxyl species start, and OH^- occupies active sites of the adsorbent. Similar results have been reported by other researchers [34, 46–48].

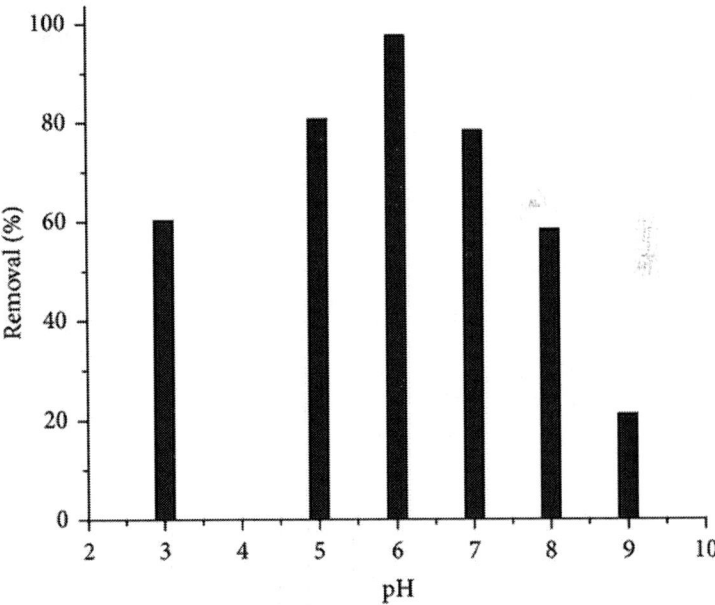

Figure 3: Effect of pH on the Hg (II) removal by ALP biomass.

Effect of Biomass Concentration

Amount of biomass used for the treatment studies is an important parameter, which determines the potential of adsorbent to remove mercury at a given initial concentration. The results clearly indicate the increase in Hg(II) uptake with increase in the biomass dosage from 0.25 to 1 g and accomplished the equilibrium at 40 min of contact time. Removal of Hg(II) was found to increase proportionality with the amount of ALP biomass dose until reaching a constant (Figure 4). The increase in percentage removal of Hg(II) is expected with increase in adsorbent dosage as the number of active sites increases. Hence, higher dosage of adsorbent has positive effect on the initial rate of metal ion removal. However, increase in adsorbent dose at constant metal concentration and volume will lead to unsaturation of sorption sites, and metals ions are inadequate to cover all the redeemable sites [49, 50].

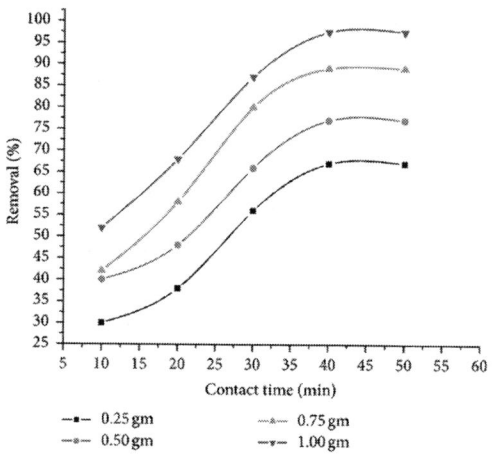

Figure 4: Effect of adsorbent dose with time on the removal of Hg(II) by ALP biomass.

At very low biomass concentration, the adsorbent surface becomes saturated with the metal ions, and the residual metal ion concentration in the solution is large. At 1 g/100 mL of biomass dosage, the Hg(II) uptake was found to be 97.5%, and this dose was taken as the optimum for further experiments.

Effect of Initial Concentration with Time

The rate of sorption is one of the most important parameters when designing a batch sorption experiment. The experimental runs measuring the effect of contact time on the biosorption of mercury at the different metal ion concentrations, 1.0 gm adsorbent dose, pH 6 and at 303 K. As shown in Figure 5, the biosorption of mercury was fast in the early stages, and the equilibrium adsorption was attained in 40 min of contact time. The observed fast biosorption kinetics is consistent with the biosorption of metal involving nonenergy-mediated reactions, where metal removal from solutions is purely due to physicochemical interactions between the biomass and the metal solution. The removal of Hg(II) increased from 23.7 mg/g to 65.1 mg/g sharply with time in the initial stage of 0–40 min range and then steady augmentation to attain equilibrium in just about 40 min time. Therefore, the optimum time and initial concentration for attaining the adsorption equilibrium is 40 min and 100 mg/L, respectively. It is perceived from the outcome that additional increase in the contact time has negligible effect on the sorption of metal ion. Many researchers have practiced the similar observation that Hg(II) removal increased almost linearly with the enhancement of the Hg(II) concentration [22, 34, 51, 52].

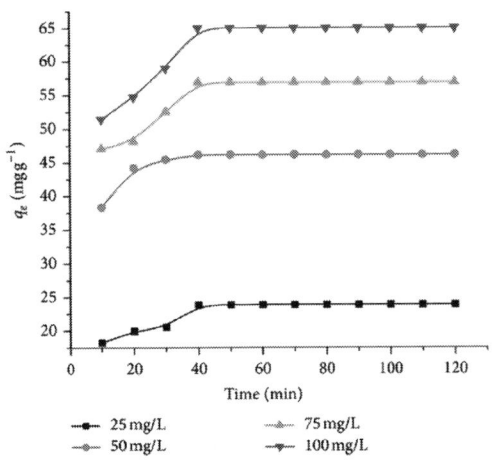

Figure 5: Effect of initial concentration and time on the sorption of Hg(II) at pH 6.0 and 1.0 g of Adulsa leaves powder.

Adsorption Isotherms

The adsorption capacity and affinity of ALP for Hg(II)) was determined with three isotherms models, namely, Langmuir, Freundlich and Temkin. The Langmuir model represents one of the first theoretical treatments of nonlinear sorption and suggests that uptake occurs on a homogeneous surface by monolayer sorption without interaction between adsorbed molecules. In addition, the model assumes uniform energies of adsorption onto the surface and no transmigration of the adsorbate. The Langmuir isotherm is represented in the following equation [33]:

$$q_e = \frac{q_m K_L C_e}{1 + K_L C_e}.$$

(2)

Equation (2) is usually linearized to obtain the following form:

$$\frac{C_e}{q_e} = \frac{1}{K_L q_m} + \left(\frac{1}{q_m}\right) C_e,$$

(3)

where C_e is the equilibrium concentration (mg·L^{-1}), q_e is the amount of adsorbed species per gram of adsorbent (mg·g^{-1}), K_L is the Langmuir equilibriumconstant related to the energy or net enthalpy of adsorption, and q_m (mg·g^{-1}) is the amount of adsorbate required to complete monolayer coverage. The plot of C_e/q_e versus C_e was analyzed to find out the Langmuir isotherm parameters which are given in Table 1. From the results, it is shown that Langmuir plot (Figure 6) gives a good fit to the experimental data with coefficient of determination $R^2 = 0.9909$. The maximum biosorption capacity of ALP biomass for Hg(II) was found to be 107.5mg g^{-1}. The value of adsorption energy, K_L was found to be 0.0913 Lmg^{-1}.

Table 1: Isotherms parameters for adsorption of Hg(II) onto ALP biomass at 303 K

Langmuir model	$\dfrac{C_e}{q_e} = \dfrac{1}{K_L q_m} + \left(\dfrac{1}{q_m}\right) C_e$	K_L (L/mg)	0.0913
		q_m (mg/g)	107.5
		R^2	0.990
Freundlich model	$\ln q_e = \left(\dfrac{1}{n}\right) \ln C_e + \ln K_F$	K_F((mg/g)/mg/ L)1/n)	1.32
		n	2.076
		R^2	0.940
Temkin model	$q_e = B \ln A + B \ln C_e$ $B = \dfrac{RT}{b}$	A(L g)	7.39
		B(KJ/mol)	0.2142
		R^2	0.973

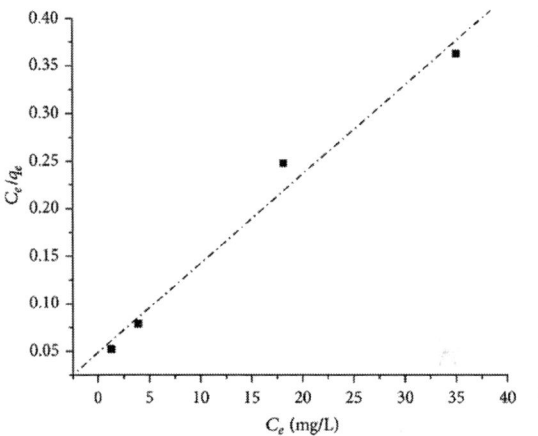

Figure 6: Langmuir isotherm model of Hg (II) sorption onto ALP biomass (at 303 K, pH 6.0).

The shape of the Langmuir isotherm can be expressed in terms of a dimensionless constant called separation factor or equilibrium parameter (R_L), which is represented as

$$R_L = \frac{1}{1 + K_L C_o},$$

(4)

where C_0 (mg·L^{-1}) is the initial concentration of the metal ion. If the average of the R_L values from the different initial concentrations used is between 0 and 1 ($0 < R_L < 1$), it indicates favorable adsorption process; however, a $R_L > 1$ represents an unfavorable process. Alternatively, if $R_L = 1$, adsorption is linear. Lastly, if $R_L = 0$, the adsorption process is irreversible [53]. Figure 7 shows that $0 < R_L < 0.05$ adsorption of Hg(II) onto ALP biomass. Because R_L is larger than zero, the adsorption process is considered favorable.

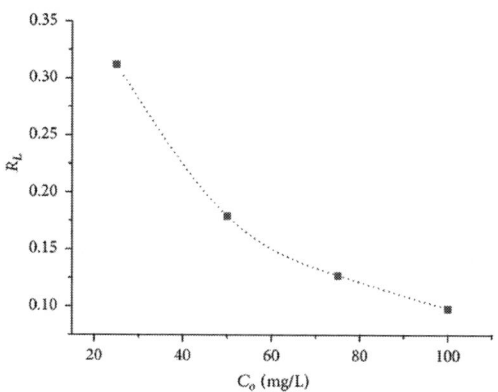

Figure 7: Separation factor for adsorption of Hg(II) by ALP biomass.

The Freundlich isotherm is a nonlinear sorption model. This model proposes a monolayer sorption with a heterogeneous energetic distribution of active sites, accompanied by interactions between adsorbed molecules. The general form of this model is [54]

$$q_e = K_F C_e^{1/n},$$

$$\ln q_e = \left(\frac{1}{n}\right) \ln C_e + \ln K_F,$$

(5)

where K_F and n are the Freundlich equilibrium constants related to the adsorption capacity and intensity of adsorption, respectively. The values of K_F and n were determined from a plot of $\ln(q_e)$ versus $\ln(C_e)$ as shown in Figure 8. The K_F constant in the Freundlich equilibrium was found to be 1.32 (mg/g)/ mg/L)$^{1/n}$. The value of n was between 0 and 10, suggesting relatively strong adsorption of these ions onto the surface of ALP biomass; for this study, we found a value of 2.076 for n. However, low correlation coefficients ($R^2 = 0.940$) suggest that this was not the best model to describe these equilibria. Similar result for the magnitude of n was described by several researchers [55–57].

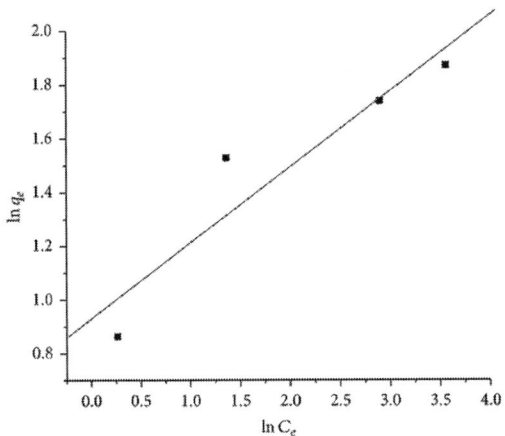

Figure 8: Freundlich isotherm model of Hg (II) sorption onto ALP biomass (at 303 K, pH 6.0).

Temkin isotherm model considered the effects of indirect adsorbate-adsorbate interaction isotherms which explained that the heat of adsorption of all the molecules on the adsorbent surface layer would decrease linearly with coverage due to adsorbate-adsorbate interactions. Therefore, the adsorption potentials of the adsorbent for the adsorbate can be evaluated using Temkin adsorption isotherm model, which assumes that the fall in the heat of sorption is linear rather than logarithmic as implied in the Freundlich equation [58]. The Temkin isotherm can be given as

$$q_e = \left(\frac{RT}{b} \right) \text{In } AC_e.$$

(6)

Equation (6) can be expressed in its linear form as

$$q_e = B \ln A + B \ln C_e,$$

(7)

With

$$B = \frac{RT}{b},$$

(8)

Where A is the equilibriumbinding constant (L·g⁻¹), b (J/mol) is a constant related to heat of adsorption, is the gas constant (8.314 J/mol/K), and T is the absolute temperature (K). As shown in Figure 9, the plot of q_e versus ln C_e enables to determine the isotherm constants A, b from the slope and intercept, respectively. The correlation coefficient $R^2 = 0.973$ for the adsorption of Hg(II) ion in temkin isotherm was fairly fitted well as compared to the freundlich isotherm. The experimental and theoretical adsorption capacity was calculated from isotherm models. The calculated parameters for all the isotherms are accessible in Table 1. The bestfit experimental equilibrium data derived from the Langmuir model suggested that the monolayer coverage and chemisorption of Hg(II) onto ALP.

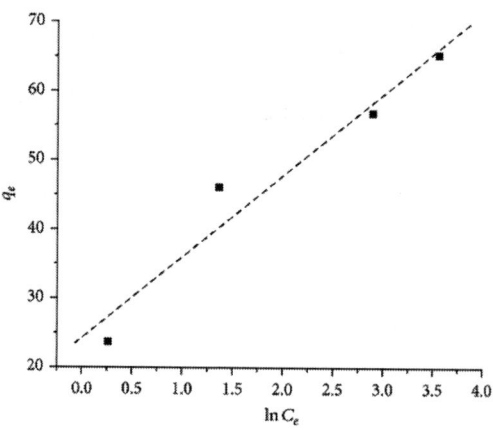

Figure 9: Temkin isotherm model of Hg (II) sorption onto ALP biomass (at 303 K, pH 6.0).

Biosorption Kinetics

Lagergren's pseudofirst-order [59] and Ho's pseudosecond-order models [60] were applied to the experimental data in order to clarify the biosorption kinetics of Hg(II) ions onto ALP. The expression for the pseudofirst-order by Lagergren is given by the differential rate law:

$$\frac{dq_t}{dt} = k_1 \left(q_e - q_t \right),$$

(9)

Where q_e is the amount of solute adsorbed at equilibrium per unit weight of adsorbent (mg g^{-1}), q_t (mg g^{-1}) is the amount of the metal ion bio adsorbed at equilibrium and time t, and k_1 (min^{-1}) is the adsorption constant. Equation (9) was integrated under the boundary conditions, giving a linear expression:

$$\ln \left(q_e - q_t \right) = \ln q_e - k_1 t.$$

(10)

The linear plot of $\ln(q_e - q_t)$ versus t shows the applicability of Lagergren equation which is shown in Figure 10. The value of k_1 and q_e calculated from the linear pseudofirstorder kinetic model and the corresponding correlation coefficients (R^2) are summarized in Table 2. The correlation coefficient for the pseudofirst-order kinetic model obtained at the studied optimum condition was 0.983.

Table 2: Kinetic parameters for adsorption of Hg(II) onto ALP biomass at 303 K

Kinetic model	Linear equation	Reaction constant	q_e(mg/g)	R^2
Pseudofirst order	$\ln(q_e - q_t) = \ln q_e - k_1 t$	$K_1 =$ 0.1209 min^{-1}	32.1	0.983
Pseudosecond order	$\dfrac{t}{q_t} = \dfrac{1}{(k_2 q_e^2)} + \left(\dfrac{1}{q_e}\right) t$	$K_2 = 2 \times 10^{-3}$ g/mg min $h = 4.34$ mg/g·min	46.6	0.998
Elovich	$q_t = \beta \ln (\alpha\beta) + \ln(t)$	$B = 5.86$ g/mg $A = 38$ mg/g·min	—	0.9661

| Intraparticle diffusion | $q_t = K_{id}t^{1/2}$ | $K_{id} = 1.70$ mg/g·min$^{0.5}$ | — | 0.9152 |
| | | $I = 3.1029$ | | |

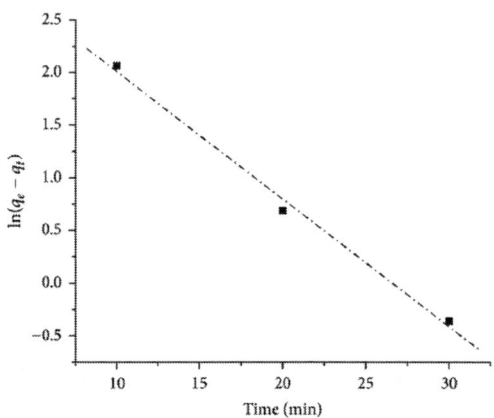

Figure 10: Lagergren plot for the adsorption of Hg(II) ions by ALP biomass at pH 6.

The expression for the pseudosecond-order model is given by the differential rate law:

$$\frac{dq_t}{dt} = k_2 (q_e - q_t)^2$$

(11)

which is on integration under the boundary conditions of $t = 0$ to $t > 0$ and $q_t = 0$ to $q_t > 0$, and after rearranging (11), the following linearized form of the pseudosecond-order model was obtained:

$$\frac{t}{q_t} = \frac{1}{(k_2 q_e^2)} + \left(\frac{1}{q_e}\right) t.$$

(12)

The initial adsorption rate (h) can be determined from k_2 and q_e values using

$$h = k_2 q_e^2,$$

(13)

Where k_2 is the rate constant of the pseudosecond-order sorption (gmg^{-1} min^{-1}) which can be obtained by plot of t/q_t against t The rate constant k_2 and equilibrium amount of metal ion q_e can be determined from slope and intercept of the plot (Figure 11). The values of k_2, q_e and the initial adsorption rate (h) were calculated as 2×10^{-3} gmg^{-1} min^{-1}, 46.6mg g-1, and 4.34mgg-1 min-1, respectively. The correlation coefficient for the pseudosecond order was 0.998.The excellent linearity and high value of correlation coefficient (R^2) from Figure 11 shows that the process follows the pseudosecond-order model with good fit in comparison to pseudofirst order model. This suggests that sorption of the metal ions involve two species in this case, the metal ion and the biomass. These results are in accordance with similar works on other metal ions [61–63] with several other natural sorbents.

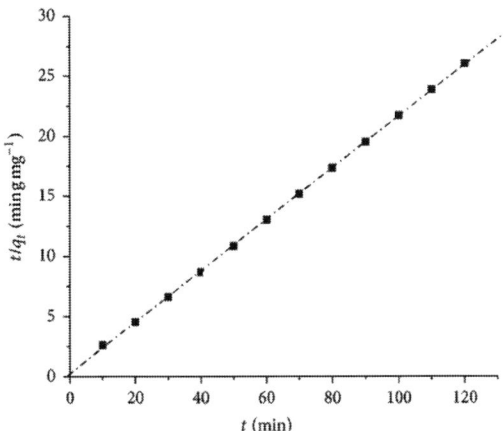

Figure 11: Pseudosecond-order kinetics for the adsorption of Hg (II) by ALP biomass at pH 6.

Elovich Kinetic Model

The Elovich kinetic model is for general application to chemisorption kinetics. The general explanation for this form of kinetic law involves that the active sites are heterogeneous in nature and therefore exhibit different activation energies for chemisorptions. The Elovich model can be expressed in the following form [64]:

$$\frac{dq_t}{dt} = \propto \exp\left(-\beta q_t\right).$$

(14)

To simplify the above equation, assume $\alpha\beta \gg 1$ [65], and by applying the boundary conditions $q_t = 0$ at $t = 0$ and $q_t = q_t$ at $t = t$, (14) becomes [66]

$$q_t = \beta \ln\left(\propto \beta\right) + \ln t,$$

(15)

where q_t is the adsorption capacity at time t (mg·g^{-1}), α is the initial adsorption rate (mg·g^{-1} min^{-1}) and β is the constant (g·mg^{-1}), which are obtained from the intercept and the slope of a plot of q_t versus $\ln t$. The plot should give a linear relationship for the applicability of simple Elovich kinetics as in Figure 12. The correlation coefficient R^2 is obtained as 0.9661 for Hg(II) ions which is found to be less than the values calculated using pseudofirst-order kinetic model and pseudosecond-order kinetic model as shown in Table 2. The calculated reaction constants α and β were found 5.86 g mg^{-1} and 38 mg g^{-1} min^{-1}, respectively.

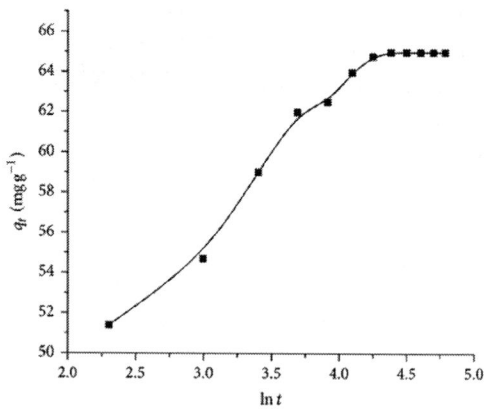

Figure 12: Elovich kinetic model for the adsorption of Hg(II) onto ALP biomass.

Intraparticle Diffusion

The adsorbate transport from the solution phase to the surface of the adsorbent particles occurs in several steps. The overall adsorption process may be controlled either by one or more steps, for example, film or external diffusion, pore diffusion, surface diffusion, and the adsorption on the pore surface or a combination of more than one steps. The adsorption rate parameter which controls the batch process for most of the contact time is the intraparticle diffusion. Good linearization of the data is observed for the initial phase of the reaction in accordance with the expected behavior if intraparticle diffusion is the rate-limiting step [67]. The possibility of intraparticle diffusion was explored by using Weber and Morris equation:

$$q_t = K_{id}t^{1/2} + I.$$

(16)

The slope and intercept of plot q_t versus $t^{1/2}$ were used to calculate the intraparticle diffusion rate constant, K_{id} (mg g^{-1} min$^{0.5}$). Values of I give an idea about the thickness of the boundary layer; that is, the larger the intercept, the greater is the boundary layer effect. The deviation of straight lines from the origin, as shown in the Figure 13, may be because of the difference between the rate of mass transfer in the initial and final steps of adsorption. Further, such deviation of straight line from the origin indicates that the pore diffusion is not the sole rate-controlling step [68]. It can also be concluded on the basis of I value as $I \neq 0$ thus suggesting that intraparticle diffusion is not the rate-limiting step. The correlation coefficient ($R^2 = 0.9152$) value was calculated from the respective plot and provided in Table 2. The value of the correlation coefficient was not uniform or large enough to suggest that intraparticle diffusion is the rate determining step of the adsorption process. The intraparticle diffusion rate constant (K_{id}) and thickness of the boundary layer (I) values were 1.70 mg g^{-1} min$^{0.5}$ and 3.1029, respectively.

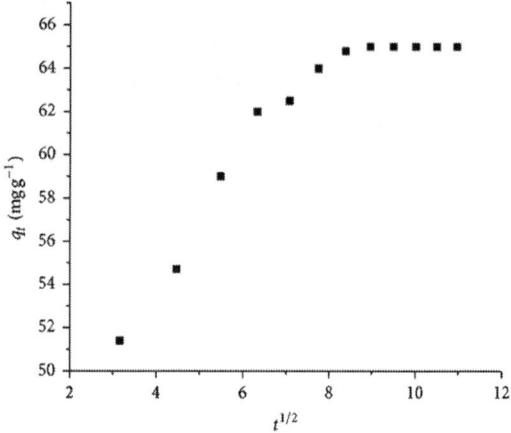

Figure 13: Intraparticle diffusion of Hg(II) adsorption onto ALP biomass.

COMPARISON OF THE PRESENT STUDY WITH LITERATURE

Table 3 shows various adsorbents previously studied for Hg(II) removal. Although the data collected in this table may or may not represents equivalent or optimized conditions or various mercury removal mechanisms in each case, it still provides a useful comparison regarding decision of selection of suitable adsorbent. The maximum adsorption capacity of mercury by ALP biomass in this study is comparable with these data. Indeed mercury adsorption by ALP biomass in this study was significantly higher than most of the selected biomass.

Table 3: Comparison of monolayer maximum adsorption capacity of Hg(II) ion with various adsorbents

Biosorbents	q_m(mg g^{-1})	References
Wheat bran	70.0	[21]
Waste rubber	4.0	[24]
Eucalyptus bark	33.1	[18]
Coir pith	154	[31]

Furfural	174	[33]
Walnut shell	151.51	[22]
Peanut hull	110	[28]
Carica papaya	155.6	[19]
Sago waste	55.6	[32]
Fruit shell of Terminalia catappa	94.4	[34]
Adulsa (Justicia Adhatoda) leaves powder	107.5	Present study
Polyaniline/attapulgite	800	[35]
Moss	94.4	[36]
Malt spent rootless	50	[37]
Acrylic textile fibre	290–710	[38]
Potamogeton natans	180	[39]
Macroalgae	329	[40]
Fertilizer waste	3.62×10^{-3}	[41]
Camel bone charcoal	28.2	[42]
Garlic (Allium sativum L.)	0.6497	[43]
Carbon aerogel	34.96	[44]

CONCLUSIONS

The study shows that Adulsa leaves can be used as a sorbent for the removal of Hg(II) ions from aqueous media. FTIR data confirmed that the functional groups such as hydroxyl, carboxyl, and amine groups were responsible for the adsorption of Hg(II) ion onto ALP biomass. Under batch condition, equilibrium was attained within 40 min. The amount of metal removal at equilibrium (40 min) increases from 67 to 97.5% with an increase of adsorbent dosage between 0.25 and 1 gm. On changing the initial concentration from 25 to 100 mg/L, the amount of mercury adsorbed increased from 23.7 to 65 mg/g at 303 K for a period of 40 min. The Langmuir isotherm that described the adsorption of Hg (II) ions onto the ALP biomass ($R^2 = 0.990$) was better than the freundlich model ($R^2 = 0.940$) and the temkin model ($R^2 = 0.973$). The study demonstrated that under optimum conditions (pH = 6.0, biomass dosage = 1 gm, temperature = 303 K, and contact time = 40 min), maximum adsorption capacity for Hg (II) was found to be 107.5 mg/g from Langmuir isotherm. Pseudosecond-order kinetics explained the adsorption of metal ion better than the pseudofirst order. K_{id} indicates that the pore diffusion is not the sole rate-controlling step. Furthermore, it can be concluded that Adulsa leaves powder hold great potential to be an environmentally friendly effective adsorbent for the removal of mercury ions from contaminated waters.

REFERENCES

1. F. M. M. Morel, A. M. L. Kraepiel, and M. Amyot, "The chemical cycle and bioaccumulation of mercury," Annual Review of Ecology and Systematics, vol. 29, pp. 543–566, 1998. · ·

2. R. Baeyens, R. Ebinghous, and O. Vasilev, Global and Regional Mercury Cycles: Sources, Fluxes and Mass Balances, Kluwer Academic, 1996.

3. D. W. Boening, "Ecological effects, transport, and fate of mercury: a general review," Chemosphere, vol. 40, no. 12, pp. 1335–1351, 2000. · ·

4. N. Pirrone, S. Cinnirella, X. Feng et al., "Global mercury emissions to the atmosphere from anthropogenic and natural sources," Atmospheric Chemistry and Physics, vol. 10, no. 13, pp. 5951–5964, 2010. · ·

5. S. Innanen, "The ratio of anthropogenic to natural mercury release in Ontario: three emission scenarios," Science of the Total Environment, vol. 213, no. 1–3, pp. 25–32, 1998. · ·

6. T. Morimoto, S. Wu, M. Azhar Uddin, and E. Sasaoka, "Characteristics of the mercury vapor removal from coal combustion flue gas by activated carbon using H_2S," Fuel, vol. 84, no. 14-15, pp. 1968–1974, 2005. · ·

7. Y. H. Li, C. W. Lee, and B. K. Gullett, "Importance of activated carbon›s oxygen surface functional groups on elemental mercury adsorption," Fuel, vol. 82, no. 4, pp. 451–457, 2003. · ·

8. F. Di Natale, A. Lancia, A. Molino, M. Di Natale, D. Karatza, and D. Musmarra, "Capture of mercury ions by natural and industrial materials," Journal of Hazardous Materials, vol. 132, no. 2-3, pp. 220–225, 2006. · ·

9. WHO, UNEP, "Guidance for Identifying Populations at Risk from Mercury Exposure, UNEP DTIE Chemicals Branch and WHO Department of Food Safety," Zoonoses and Foodborne Diseases, p. 4, 2008.

10. R. H. Crist, J. R. Martin, J. Chonko, and D. R. Crist, "Uptake of metals on peat moss: an ion-exchange process," Environmental Science and Technology, vol. 30, no. 8, pp. 2456–2461, 1996.

11. C. P. Huang and D. W. Blankenship, "The removal of mercury (II) from dilute aqueous solution by activated carbon," Water Research, vol. 18, no. 1, pp. 37–46, 1984.

12. H. K. Alluri, S. R. Ronda, V. S. Settalluri, B. Jayakumar Singh, V. Suryanarayana, and P. Venkateshwar, "Biosorption: an eco-friendly alternative for heavy metal removal," African Journal of Biotechnology, vol. 6, no. 25, pp. 2924–2931, 2007.

13. B. Volesky and Z. R. Holan, "Biosorption of heavy metals," Biotechnology Progress, vol. 11, no. 3, pp. 235–250, 1995.

14. B. Volesky, "Detoxification of metal-bearing effluents: biosorption for the next century," Hydrometallurgy, vol. 59, no. 2-3, pp. 203–216, 2001.

15. B. S. Marina, M. P. Jelena, and N. R. Radojka, "Biosorption of copper(ii) and chromium(vi) by modified tea Fungus," APTEFF, vol. 43, pp. 1–342, 2012.

16. Y. Nuhoglu and E. Malkoc, "Thermodynamic and kinetic studies for environmentaly friendly Ni(II) biosorption using waste pomace of olive oil factory," Bioresource Technology, vol. 100, no. 8, pp. 2375–2380, 2009.

17. A. Khanafari, S. Eshghdoost, and A. Mashinchian, "Removal of lead and chromium from aqueous solution by Bacillus circulans biofilm," Iranian Journal of Environmental Health Science and Engineering, vol. 5, no. 3, pp. 195–200, 2008.

18. I. Ghodbane and O. Hamdaoui, "Removal of mercury(II) from aqueous media using eucalyptus bark: kinetic and equilibrium studies," Journal of Hazardous Materials, vol. 160, no. 2-3, pp. 301–309, 2008.

19. S. Basha, Z. V. P. Murthy, and B. Jha, "Sorption of Hg(II) onto Carica papaya: experimental studies and design of batch sorber,"

Chemical Engineering Journal, vol. 147, no. 2-3, pp. 226–234, 2009. · ·

20. F. Rozada, M. Otero, A. Morán, and A. I. García, "Adsorption of heavy metals onto sewage sludge-derived materials," Bioresource Technology, vol. 99, no. 14, pp. 6332–6338, 2008. · ·

21. M. A. Farajzadeh and A. B. Monji, "Adsorption characteristics of wheat bran towards heavy metal cations," Separation and Purification Technology, vol. 38, no. 3, pp. 197–207, 2004. · ·

22. M. Zabihi, A. Ahmadpour, and A. Haghighi Asl, "Removal of mercury from water by carbonaceous sorbents derived from walnut shell," Journal of Hazardous Materials, vol. 167, no. 1–3, pp. 230–236, 2009. · ·

23. G. Macchi, D. Marani, and G. Tiravanti, "Uptake of mercury by exhausted coffee grounds,"Environmental Technology Letters, vol. 7, no. 8, pp. 431–444, 1986.

24. W. R. Knocke and L. H. Hemphill, "Mercury(II) sorption by waste rubber," Water Research, vol. 15, no. 2, pp. 275–282, 1981. · ·

25. M. K. Sreedhar, A. Madhukumar, and T. S. Anirudhan, "Evaluation of an adsorbent prepared by treating coconut husk with polysulphide for the removal of mercury from wastewater," Indian Journal of Engineering and Materials Sciences, vol. 6, no. 5, pp. 279–285, 1999.

26. S. K. Srivastava, R. Tyagi, and N. Pant, "Adsorption of heavy metal ions on carbonaceous material developed from the waste slurry generated in local fertilizer plants," Water Research, vol. 23, no. 9, pp. 1161–1165, 1989.

27. B. Volesky, "Removal and recovery of heavy metals by biosorption," in Biosorption of Heavy Metals, B. Volesky, Ed., pp. 7–43, CRC Press, Boca Raton, Fla, USA, 1990.

28. C. Namasivayam and K. Periasamy, "Bicarbonate-treated peanut hull carbon for mercury (II) removal from aqueous solution," Water Research, vol. 27, no. 11, pp. 1663–1668, 1993. · ·

29. B. S. Inbaraj and N. S. Sulochana, "Utilization of an agricultural waste jack fruit peel for the removal of Hg(II) from aqueous solution," in Proceeding of 17th International Conference on Solid Waste Technology and Management, R. L. Mersky, Ed., pp. 802–811, Philadelphia, Pa, USA, 2001.

30. A. K. Sen and A. K. De, "Adsorption of mercury(II) by coal fly ash," Water Research, vol. 21, no. 8, pp. 885–888, 1987.

31. C. Namasivayam and K. Kadirvelu, "Uptake of mercury (II) from wastewater by activated carbon from an unwanted agricultural solid by-product: coirpith," Carbon, vol. 37, no. 1, pp. 79–84, 1999.

32. K. Kadirvelu, M. Kavipriya, C. Karthika, N. Vennilamani, and S. Pattabhi, "Mercury (II) adsorption by activated carbon made from sago waste," Carbon, vol. 42, no. 4, pp. 745–752, 2004. · ·

33. I. Langmuir, "The constitution and fundamental properties of solids and liquids. Part I. Solids," The Journal of the American Chemical Society, vol. 38, no. 2, pp. 2221–2295, 1916.

34. B. S. Inbaraj and N. Sulochana, "Mercury adsorption on a carbon sorbent derived from fruit shell ofTerminalia catappa," Journal of Hazardous Materials, vol. 133, no. 1–3, pp. 283–290, 2006. · ·

35. H. Cui, Y. Qian, Q. Li, Q. Zhang, and J. Zhai, "Adsorption of aqueous Hg(II) by a polyaniline/attapulgite composite," Chemical Engineering Journal, vol. 211-212, pp. 216–223, 2012.

36. A. Sari and M. Tuzen, "Removal of mercury(II) from aqueous solution using moss (Drepanocladus revolvens) biomass: equilibrium, thermodynamic and kinetic studies," Journal of Hazardous Materials, vol. 171, no. 1–3, pp. 500–507, 2009. · ·

37. V. A. Anagnostopoulos, I. D. Manariotis, H. K. Karapanagioti, and C. V. Chrysikopoulos, "Removal of mercury from aqueous solutions by malt spent rootlets," Chemical Engineering Journal, vol. 213, pp. 135–141, 2012.

38. J. V. Nabais, P. J. M. Carrott, M. M. L. R. Carrott et al., "Mercury removal from aqueous solution and flue gas by adsorption on activated carbon fibres," Applied Surface Science, vol. 252, no. 17, pp. 6046–6052, 2006. · ·

39. C. Lacher and R. W. Smith, "Sorption of Hg(II) by Potamogeton natans dead biomass," Minerals Engineering, vol. 15, no. 3, pp. 187–191, 2002. · ·

40. R. Herrero, P. Lodeiro, C. Rey-Castro, T. Vilariño, and M. E. Sastre De Vicente, "Removal of inorganic mercury from aqueous solutions by biomass of the marine macroalga Cystoseira baccata," Water Research, vol. 39, no. 14, pp. 3199–3210, 2005. · ·

41. D. Mohan, V. K. Gupta, S. K. Srivastava, and S. Chander, "Kinetics of mercury adsorption from wastewater using activated carbon derived from fertilizer waste," Colloids and Surfaces A, vol. 177, no. 2-3, pp. 169–181, 2001. · ·

42. S. S. M. Hassan, N. S. Awwad, and A. H. A. Aboterika, "Removal of mercury(II) from wastewater using camel bone charcoal," Journal of Hazardous Materials, vol. 154, no. 1–3, pp. 992–997, 2008. · ·

43. Y. Eom, J. H. Won, J.-Y. Ryu, and T. G. Lee, "Biosorption of mercury(II) ions from aqueous solution by garlic (Allium sativum L.) powder," Korean Journal of Chemical Engineering, vol. 28, no. 6, pp. 1439–1443, 2011. · ·

44. K. Kadirvelu, J. Goel, and C. Rajagopal, "Sorption of lead, mercury and cadmium ions in multi-component system using carbon aerogel as adsorbent," Journal of Hazardous Materials, vol. 153, no. 1-2, pp. 502–507, 2008. · ·

45. S. Schiewer and B. Volesky, "Biosorption process for heavy metal removal," in Environmental Microbe-Metal Interactions, D. R. Lovley, Ed., pp. 329–362, ASM Press, Washington, DC, USA, 2000.

46. M. Amini, H. Younesi, N. Bahramifar et al., "Application of response surface methodology for optimization of lead biosorption in an aqueous solution by Aspergillus niger," Journal of Hazardous Materials, vol. 154, no. 1–3, pp. 694–702, 2008. · ·

47. G. Bayramoğlu, I. Tuzun, G. Celik, M. Yilmaz, and M. Y. Arica, "Biosorption of mercury(II), cadmium(II) and lead(II) ions from aqueous system by microalgae Chlamydomonas reinhardtiiimmobilized in alginate beads," International Journal of Mineral Processing, vol. 81, no. 1, pp. 35–43, 2006. · ·

48. Y. Zeroual, A. Moutaouakkil, F. Z. Dzairi et al., "Biosorption of mercury from aqueous solution by Ulva lactuca biomass," Bioresource Technology, vol. 90, no. 3, pp. 349–351, 2003. · ·

49. R. Gong, Y. Ding, H. Liu, Q. Chen, and Z. Liu, "Lead biosorption and desorption by intact and pretreated spirulina maxima biomass," Chemosphere, vol. 58, no. 1, pp. 125–130, 2005. · ·

50. N. Saifuddin and A. Z. Raziah, "Removal of heavy metals from industrial effluent using Saccharomyces cerevisiae (Baker's yeast) immobilized in chitosan/lignosulphonate matrix," Journal of Applied Science Research, vol. 3, pp. 2091–2099, 2007.

51. F.-S. Zhang, J. O. Nriagu, and H. Itoh, "Photocatalytic removal and recovery of mercury from water using TiO_2-modified sewage sludge carbon," Journal of Photochemistry and Photobiology A, vol. 167, no. 2-3, pp. 223–228, 2004. · ·

52. M. F. Yardim, T. Budinova, E. Ekinci, N. Petrov, M. Razvigorova, and V. Minkova, "Removal of mercury (II) from aqueous solution by activated carbon obtained from furfural," Chemosphere, vol. 52, no. 5, pp. 835–841, 2003. · ·

53. W. S. Wan Ngah, A. Kamari, and Y. J. Koay, "Equilibrium and kinetics studies of adsorption of copper (II) on chitosan and chitosan/PVA beads," International Journal of Biological Macromolecules, vol. 34, no. 3, pp. 155–161, 2004. · ·

54. H. M. F. Freundlich, "Uber die adsorption in lasugen," Zeitschrift Für Physikalische Chemie (Leipzig), vol. 57A, pp. 385–470, 1906.

55. Y.-M. Hao, C. Man, and Z.-B. Hu, "Effective removal of Cu (II) ions from aqueous solution by amino-functionalized magnetic nanoparticles," Journal of Hazardous Materials, vol. 184, no. 1–3, pp. 392–399, 2010. · ·

56. T. K. Naiya, A. K. Bhattacharya, and S. K. Das, "Adsorption of Cd(II) and Pb(II) from aqueous solutions on activated alumina," Journal of Colloid and Interface Science, vol. 333, no. 1, pp. 14–26, 2009. · ·

57. J. Hu, G. Chen, and I. M. C. Lo, "Removal and recovery of Cr(VI) from wastewater by maghemite nanoparticles," Water Research, vol. 39, no. 18, pp. 4528–4536, 2005. · ·

58. M. I. Temkin and V. Pyzhev, "Kinetics of ammonia synthesis on promoted iron catalysts," Acta Physiochimica URSS, vol. 12, pp. 327–356, 1940.

59. S. Lagergren, Zur Theorie Der Sogenannten Adsorption Geloster Stoffe, vol. 24, Kungliga Sevenska Vetanskapas akademiens, Handlingar, 1898.

60. Y. S. Ho and G. McKay, "Pseudo-second order model for sorption processes," Process Biochemistry, vol. 34, no. 5, pp. 451–465, 1999. · ·

61. Y. Prasanna Kumar, P. King, and V. S. R. K. Prasad, "Equilibrium and kinetic studies for the biosorption system of copper(II) ion from aqueous solution using Tectona grandis L.f. leaves powder," Journal of Hazardous Materials, vol. 137, no. 2, pp. 1211–1217, 2006. · ·

62. A. Lodi, C. Solisio, A. Converti, and M. Del Borghi, "Cadmium, zinc, copper, silver and chromium(III) removal from wastewaters by Sphaerotilus natans," Bioprocess Engineering, vol. 19, no. 3, pp. 197–203, 1998. · ·

63. Y.-S. Ho, W.-T. Chiu, C.-S. Hsu, and C.-T. Huang, "Sorption of lead ions from aqueous solution using tree fern as a sorbent," Hydrometallurgy, vol. 73, no. 1-2, pp. 55–61, 2004. · ·

64. M. J. D. Low, "Kinetics of chemisorption of gases on solids," Chemical Reviews, vol. 60, no. 3, pp. 267–312, 1960.

65. S. H. Chien and W. R. Clayton, "Application of Elovich equation to the kinetics of phosphate release and sorption in soils," Soil Science Society of America Journal, vol. 44, pp. 265–268, 1980.

66. D. L. Sparks, Kinetics of Reaction in Pure and Mixed Systems, Soil Phy Chem CRC Press, Boca Raton, Fla, USA, 1986.

67. W. J. Weber and J. C. Morris, "Kinetics of adsorption on carbon from solution," Journal of the Sanitary Engineering Division, American Society of Civil Engineers, vol. 89, pp. 31–60, 1963.

68. S. C. Ibrahim, M. A. K. M. Hanafiah, and M. Z. A. Yahya, "Removal of Cadmium form aqueous solutions by adsorption on sugarcane bagasse," American Europian Journal of Agricultural & Environmental Sciences, vol. 1, no. 3, pp. 179–184, 2006.

Chapter 6

The Effect of Industrial Heavy Metal Pollution on Microbial Abundance and Diversity in Soils – A Review

Anna Lenart-Boroń[1] and Piotr Boroń[2]

[1]Department of Microbiology, Faculty of Agriculture and Economics, University of Agriculture in Cracow, Cracow, Poland

[2]Department of Forest Pathology, Faculty of Forestry, University of Agriculture, Cracow, Poland

INTRODUCTION

Metals are essential components of the ecosystem, whose biologically available concentrations depend mainly on geological and biological processes [1]. There are several definitions of heavy metals, and some of them are based on the mass density of these elements. Authors of numerous publications use different limits to define the threshold density for a "heavy metal", ranging from 3.5 to 7 g×cm^{-3}, however, the majority of authors suggests that the mass density of heavy metals should be greater than 4.5 g×cm^{-3} [2]. Within the group of heavy metals one can distinguish both elements that are essential for living organisms (microelements) and the elements whose physiological role is unknown and thus they are "inactive" towards organisms. The metals that serve as microelements in living organisms usually occur in trace amounts, precisely defined for each species and both their deficiency and excess badly affect living organisms [3]. The term "heavy metal" is linked in many people's minds to metals that are toxic. However, this is not always the truth. The effect of any substance on a living system is always dependent on its available concentration to cells. Also, several heavy metal ions are crucial in metabolic processes at low concentrations but are toxic at high concentrations [2]. Nevertheless, locally elevated levels of these elements can create significant environmental and health problems when the release of metals through various biological, geological and anthropogenic processes far exceeds its natural content resulting from processes of metal cycling. Heavy metal pollution of terrestrial environments is of great concern, due to the persistence of metals in the ecosystem and their threat to all living organisms [4].

Given the importance of the subject of soil heavy metal pollution and its effect on soil microorganisms, this chapter gives an overview of the severity of the problem when it comes to the reaction of soil microbial community to the environmental pollution. The first part of this chapter deals with the abundance of microorganisms in soils and their role in this environment. The next part concerns major sources of heavy metals in soils with particular emphasis on the most important source of soil pollution, i.e. human activity (and more precisely – industry and mining). The following part discusses the effects that toxic levels of heavy metals may have on the microbial population in soils. The last two parts of this chapter describe the ways of dealing with heavy metal pollution – one introduces the term of phytoremediation (soil remediation with the use of plants) and the other one focuses on the use of microorganisms resistant to heavy metals in the process of soil remediation.

THE COMPLEXITY OF MICROBIAL COMMUNITY IN SOILS

Except for occasional insects or earthworms, once visible traces of plant biomass are removed, soil appears as a lifeless mass that is composed of mineral particles and organic residues. However, even desert soils are abundant source of living microorganisms. This seemingly lifeless matter contains complex microbial community, including bacteria, fungi, protozoa and viruses. The integrity of the aboveground and belowground ecosystems depends on the stability, resilience and function of the soil microbial community [5].

Soil is an interesting medium for growing microorganisms, as it contains various nutrients that the microbes need for their metabolism. Unfortunately, nutrients are not always readily available [6]. However, it is one of the richest reservoirs of microorganisms, i.e. 1 gram of agricultural soil may contain even several billion colony forming units (CFUs) of microorganisms belonging to thousands of different species [7], and even though microorganisms constitute less than 0.5% of the soil mass, they have a major impact on soil properties and processes [5]. Table 1 presents the average numbers of soil microorganisms in a "typical" temperate soil. Destruction of the soil microbiota through mismanagement or environmental pollution causes decline or even death of the aboveground plant and animal populations.

Table 1: Relative numbers and approximate biomass of the soil microbiota in a fertile soil [8]

Organisms	Numbers		Biomass [wet kg×ha^{-1}]
	Per m2	Per g	
Bacteria	10^{13}-10^{14}	10^8-10^9	300-3000
Actinomycetes	10^{12}-10^{13}	10^7-10^8	300-3000
Fungi	10^{10}-10^{11}	10^5-10^6	500-5000
Microalgae	10^9-10^{10}	10^3-10^6	10-1500

The most characteristic feature of microbial habitats is the great micro-spatial variability of environmental parameters, like temperature or nutrient availability. Many basic requirements of heterogeneous microorganisms are satisfied by various soil microhabitats. This is the reason why, in ecological terms, a number of varying microbial niches can be described. Therefore, the microbial community is composed of diverse taxa with different nutritional demands within small microenvironments [9]. Analysis of the spatial distribution of bacteria at microhabitat levels showed that in soils subjected to different fertilization treatments, the majority of bacteria were located in micropores of stable soil micro-aggregates (2 – 20 μm), as they contained over 80% of cells [10]. Such microhabitats offer the most favorable conditions for microbial growth in terms of water and substrate availability, gas diffusion and protection against predation. The microhabitat-adapted groups of microorganisms form so-called consortia which are held together by mutually facilitating metabolic processes. The consortia are characterized by more or less sharp boundaries, and variable level of interaction with each other and with other parts of the soil biota. Numerous investigations emphasize the impact of soil structure and spatial isolation on microbial diversity and community structure [11]. Some studies indicate that the soil particle size affects the diversity of microorganisms and community structure to a greater extent than other factors such as bulk pH and the type or amount of available organic compounds [12]. Other investigations show that the type and amount of available organic substrates strongly affect the abundance of microbial groups and their functional diversity in soils [13]. Fierer and Jackson [14] claim that the structure of soil bacterial communities is not random also at continental scale and that the diversity and composition of soil bacterial communities at large spatial scales can be predicted to a large extent by a single variable, that is soil pH. The diversity of soil microorganisms comprises different levels of biological organization. It includes genetic variability among taxa (species), number (richness), relative abundance (evenness) of taxa and functional groups within communities [11]. The overall biodiversity of soil microflora comprises bacteria, fungi, actinomycetes and photosynthetic microorganisms [6].

Bacteria constitute the most numerous group of soil microbes – a teaspoon of productive soil contains between 100 million and 1 billion bacterial cells. As soil environment changes rather drastically, spore-forming bacteria tend to be the most common. When environmental conditions become too difficult for normal growth, the bacteria form spores and remain dormant until the environment returns to proper conditions [6]. They facilitate various processes in soils, e.g. those related to water dynamics, nutrient cycling or disease suppression [15]. Soil-dwelling bacteria may be divided into different groups based on:

- Shapes: rods (also called bacilli), sphere (also called cocci) and spiral (also called spirilla)
- Their reaction to oxygen: aerobic (bacteria that need oxygen for their survival) and anaerobic (the ones that do not require oxygen and in most cases cannot bear oxygen that is deadly for them)
- Result of Gram staining: Gram negative (stain pink and have thinner cell walls, they are the smallest ones and tend to be more sensitive to water stress) or Gram positive (stain violet, have thicker cell walls, are larger in size and tend to resist water stress)
- Source of carbon they use: autotrophs (obtain carbon from carbon dioxide – some autotrophic bacteria directly use sunlight in order to produce sugar from carbon dioxide, while others depend on various chemical reactions) or heterotrophs (they obtain carbohydrates from their environment)
- Classification based on phyla: based on morphology, barcode DNA sequences, physiological requirements and biochemical characteristics, bacteria have been classified into 12 phyla. Each phylum corresponds to a number of bacterial species and genera [15].

Tate [5] lists the most commonly encountered soil bacterial genera as: *Acinetobacter, Agrobacterium, Alcaligenes, Arthrobacter, Bacillus, Brevibacterium, Caulobacter, Cellulomonas, Clostridium, Corynebacterium, Flavobacterium, Hyphomicrobium, Metallogenium, Micrococcus, Mycobacterium, Pseudomonas, Sarcina, Streptococcus* and *Xanthomonas*. These are the heterotrophic bacteria that are augmented in soil by autotrophic and mixotrophic representatives, including nitrifiers, *Thiobacillus*species and iron bacteria.

Bacteria facilitate a number of physical and biochemical alterations or reactions in soils and thereby directly or indirectly support the development of higher plants. Their performance is vital for a variety of processes that include: decomposition of cellulose or other carbohydrates (e.g. *Bacillus,Achromobacter, Cellulomonas, Clostridium, Methanococcus*), ammonification (*Bacillus,Pseudomonas*), nitrification (*Nitrosomonas, Nitrobacter*), denitrification (*Achromobacter,Pseudomonas, Bacillus, Micrococcus*) and nitrogen fixation (symbiotic *Rhizobium, Bradyrhizobium*etc., non-symbiotic *Azotobacter, Beijerinckia*) [16].

On the other hand, soil fungi form three functional groups: decomposers, mutualists and pathogens. Fungi, along with bacteria, are important decomposers of hard to digest organic matter and they increase nutrient uptake of phosphorus. Mycorrhizal fungi support plants by promoting root branching and increasing nitrogen, phosphorus and water uptake. They improve plant resilience to pests, diseases or drought and improve soil structure, as fungal hyphae binds soil particles together to create water-stable aggregates. They in turn create the pore spaces in the soil that enhance water retention and drainage [17]. The most common fungi found in soil belong to the *Penicillium* and *Aspergillus* genera together with the representatives of the Zygomycetes and the mycorrhizae-associated Ascomycetes and Basidiomycetes [5].

Actinomycetes are a large group of microorganisms, systematically identified as bacteria, that grow as hyphae. They decompose a wide range of substances, but they are particularly important in degrading recalcitrant (difficult to degrade) compounds such as chitin, lignin, keratin and cellulose. Moreover, they produce a number of secondary metabolites such as antibiotics i.e. streptomycin [18] or geosmine which is responsible for "earthy" smell after soil plowing [15]. Actinomycetes are important in forming stable humus, which enhances soil structure, improves soil nutrient storage and increases water retention in soils. According to Tate [5], the most commonly encountered soil actinomycetes belong to *Nocardia* and *Streptomyces* genera.

Algae are the most common among photosynthetic microorganisms found in soil. They are found only near soil surface, where light is readily available [6]. The most common genera of green algae found in soil are: *Chlorella, Chlamydomonas, Chlorococcum, Protosiphon* etc. and that of diatoms are *Navicula, Pinnularia. Synedra, Frangilaria*. Their functions include the maintenance of soil fertility, increasing water retention capacity of soil, prevention of soil erosion due to the fact that they act as cementing agents in binding soil particles. They add organic matter to soil after the cell death and thus increase the amount of organic carbon, while their photosynthetic activity release large quantity of oxygen that facilitate the aeration in submerged soils or oxygenate the soil environment. They also take part in weathering rocks, thus building up the soil structure [19]

Although biomass of all microorganisms living in soil constitutes only several percent of organic matter content, they play an important role in the functioning of entire ecosystems [20]. They take part in soil formation, mineralize organic substances, provide plants with bioavailable compounds, cooperate with plants or may be used as a source of insecticidal substances [21]. One of the most important and most widely studied microbial groups in terms of beneficial effects to soil and plants is the group of Plant Growth Promoting Rhizobacteria (PGPR) [22]. This group includes bacterial species from genera such as *Azotobacter*, *Azospirillum*, *Bacillus*, *Burkholderia*, *Enterobacter*, *Erwinia*,*Flavobacterium*, *Pseudomonas* and *Rhizobium* [23]. Activity of these bacteria significantly increases plant growth and yield due to a variety of mechanisms, such as phytohormone production, symbiotic and asymbiotic N_2 fixation, production of siderophores, activity against phytopathogenic microorganisms, synthesis of antibiotics, enzymes and/or fungicidal compounds, as well as solubilization of mineral phosphates and other nutrients [24]. PGPR may improve plant growth, salinity and metal toxicity stress tolerance, as well as they are able to produce phytohormones such as indole-3-acetic acid (IAA) [25]. Some PGPR produce the enzyme 1-aminocyclopropane-1-carboxylate (ACC) deaminase, which hydrolyses ACC, the immediate precursor of ethylene in plants. By decreasing its concentration in seedlings and thus its inhibitory effect, these PGPR stimulate seedlings' root length [26]. Figure 1. shows the ways how Plant Growth Promoting Rhizobacteria can stimulate plants. Bacteria from the genus *Rhizobium* form symbiotic associations with roots of leguminous plants like clovers, peas or alfalfa. These Gram-negative, rod-shaped bacteria infect growing root hairs, forming visible nodules. In this form of symbiosis, plants supply simple carbohydrates to bacteria while bacteria convert nitrogen (N_2) from air into the forms (NO_3^- or NH_4^+) that plant can use. When leaves or roots from the plant decompose, nitrogen content increases in soil [15]. Some microbial species are capable of detergent decomposition, taking part in self-purification process of soils. Decomposers are particularly important in immobilizing or retaining nutrients in their cells, thus preventing the loss of nutrients, such as nitrogen, from the rooting zone.

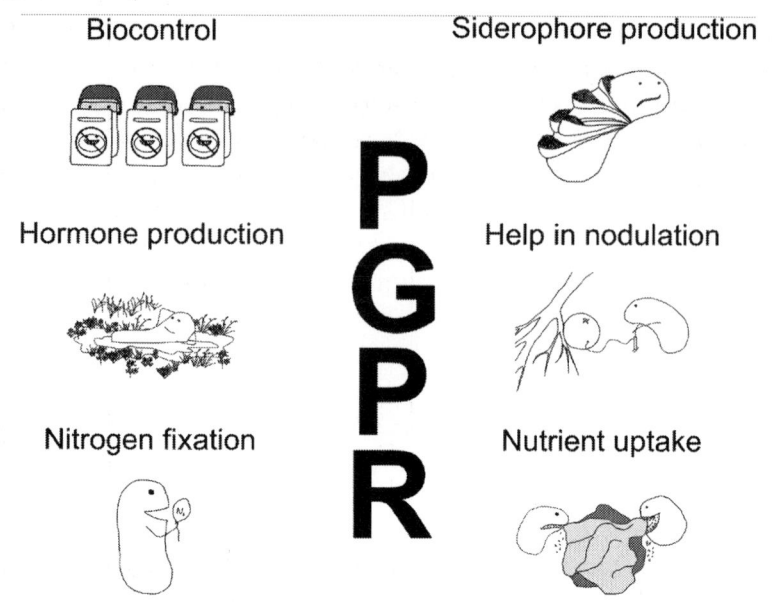

Biocontrol

Siderophore production

Hormone production

Help in nodulation

Nitrogen fixation

Nutrient uptake

P
G
P
R

Figure 1: Summary of mechanisms employed by Plant Growth Promoting Rhizobacteria to stimulate plant development.

Despite beneficial effects of numerous soil microbes on plant growth or development, soil structure and functioning, some soil-dwelling microorganisms may cause plant, animal and human diseases. Similarly to the beneficial soil microflora, soil pathogens include bacteria, fungi and viruses. One of the example of the most important or best known plant pathogens include *Agrobacterium tumefaciens* (whose updated scientific name is now *Rhizobium radiobacter*) [27] which is the causal agent of crown gall disease of walnuts, grape vines, stone fruits and many others. These bacteria infect plant roots and induce cells to divide (due to overproduction of auxin and cytokinin), causing a tumor-like swellings that contain infected cells [28]. *Erwinia carotovora* (or now called *Pectobacterium carotovorum*) and *Erwinia amylovora*, the Gram-negative plant pathogens with a diverse host range cause infections of numerous agriculturally and scientifically important plant species, such as potato, apple, pear and some members of the family *Rosaceae* [29]. Soil is also an abundant source of fungal pathogens. Among them we may distinguish *Rhizoctonia solani*, a plant pathogenic fungus with a wide host range and worldwide distribution. It causes collar rot, crown rot, root rot, damping off and wire stem [30]. It mainly attacks plant seeds below the soil surface, but may also infect leaves and stems. Due to a variety of hosts that this pathogen attacks, it is of great importance and is detrimental to a variety of crops. The *Armillaria* root rot, caused by several species of basidoimycete genus *Armillaria* – the honey fungus is, on the other hand, one of the greatest threat for woody plants [31]. Another example of soil-borne plant pathogens is an important genus of fungi – *Fusarium*, which contains a number of, worldwide distributed, phytopathogenic species [32]. Moreover, *Fusarium* has also been more recently reported as an emerging human pathogen for immunocompromised patients [33]. *Clostridium tetani* is an example of one of the most dangerous soil-borne human pathogens. It is a tetanus-causing Gram-positive bacterium, whose transmission occurs through the contamination of wounds with soil carrying its spores [34]. Generally, soil is a typical carrier of human bacterial and fungal pathogens. Another example of them is *Bacillus anthracis*, the causative agent of anthrax, which is found worldwide in a variety of soil environments. Inhalation of *B. anthracis* spores can be fatal. Nevertheless, the incidence of both of these fatal diseases has been largely controlled in developed countries due to the development of vaccines [35].

Undoubtedly, soil is an inexhaustible reservoir of microorganisms, both beneficial and pathogenic ones. Causing the imbalance between groups of soil macro- and microorganisms may be irreversible and result in a variety of effects, sometimes unpredictable. Such imbalance may be caused by soil pollution resulting from developing industry, therefore understanding the sources and effects of industrial soil pollution is an important element in preventing the environmental degradation.

SOURCES OF SOIL HEAVY METAL POLLUTION

Chemical compounds, entering the ecosystem as a result of different human activities, may accumulate in soil and water environments. Therefore, soil may be regarded as a long-term reservoir of pollutants, from which these compounds may be introduced to food chains or groundwater [36]. Inappropriate and careless disposal of industrial waste often results in environmental pollution. The pollution includes point sources such as emission, effluents and solid discharge from industry, vehicle exhaustion and metal smelting or mining, as well as nonpoint sources (e.g. the use of pesticides or excessive use of fertilizers) [37]. Each of the sources have their own damaging effects on plant, animal and human health, but those that add heavy metals to soils are of serious concern due to the persistence of these elements in the environment. They cannot be destroyed, but are only transformed from one state to another [38].

Soil pollution may be defined as presence of xenobiotics (e.g. chemical compounds, radioactive elements) that alters the soil properties – both chemical, physical and biological. Soil pollution, including heavy metals, may be of natural origin, like volcanic eruptions, animal excrements or ore leaching. Nevertheless, human activity and mostly chemical industry, mining and metallurgy, as well as municipal management and traffic emissions are the main source of environmental pollution. Some authors also mention that waste disposal, waste incineration, fertilizer application and long-term application of wastewater in agricultural lands may result in heavy metal pollution of soils [39].

Heavy metals occur naturally in soils due to pedogenetic processes of weathering parent materials, however concentrations of these metals are regarded as trace (<1000 mg×kg^{-1}) and rarely toxic [40]. Due to the disturbance and acceleration of the natural slow geochemical cycles of metals by man, most soils of rural and urban environments accumulate one or more heavy metals above the defined background levels, high enough to cause risks to ecosystems [41]. Nevertheless, heavy metals occurring in soils from anthropogenic sources tend to be more mobile, therefore more bioavailable than pedogenic or lithogenic ones [42].

Communication routes, such as roads, railways etc., are an important source of soil pollution, especially in the case of lead and zinc. Despite restricted use of leaded gasoline adopted in most countries, lead remains one of the most serious automotive-originating metal pollutant. The areas located nearby roads, particularly in urban sites, are the most vulnerable to automotive pollution. Apart from lead and zinc, chromium, cadmium, nickel and platinum are among the pollutants emitted by combustion engine-powered vehicles [43]. Heavy metals enter the environment as a result of tire wear and damage of vehicle parts. Moreover, grease used in vehicles may also be the source of cadmium pollution along roads [44]. Nickel emission results from this metal being added in gasoline and atmospheric abrasion of nickel-containing parts of automobiles [45]. The changes in the concentrations of lead, nickel, cadmium, copper and zinc in roadside soils are frequently attributed to traffic density [46].

Standard agricultural practices are also a significant source of heavy metals in soils, as application of fertilizers and pesticides has contributed to a continuous accumulation of these elements. Heavy metals can accumulate in soils due to the application of liquid and solid manure, as well as inorganic fertilizers [47]. The application of numerous biosolids, such as livestock manures, composts and municipal sewage sludge on agricultural soils leads to the accumulation of various heavy metals, such as, Cd, Cr, Cu, Hg, Mo, Ni, and Zn [48]. Lime and superphosphate fertilizers contain not only major elements necessary for plant nutrition and growth but also trace metal impurities such as cadmium. The presence of high concentrations of Cd in some fertilizers (particularly in phosphatic fertilizers) is of most concern due to the toxicity of this metal and its ability to accumulate in soils as well as due to its bioaccumulation in plant and consequently in animal tissues [49, 50]. Additionally, copper-containing compounds have been widely used in agricultural practice as pesticides. Copper oxychloride is annually applied on vineyards as a fungicide to control a significant number of plant diseases. Inevitably, this Cu ends up in the agricultural soil and adjacent pristine natural vegetation [51]. Lead arsenate was used in fruit orchards for many years to control some of the parasitic insects. Arsenic-containing compounds were also extensively used to control pests in banana plantations in New Zealand and Australia [52]. High fertilizer applications and acid atmospheric deposition, combined with insufficient liming, may also cause a decrease in pH and thus increase heavy metal bioavailability, aggravating the problem of deteriorating food quality, metal leaching and impact on soil organisms [53]. The application of municipal wastewater or industrial waste as fertilizers and liming agents in agriculture is a separate issue. Application of this type of waste requires constant monitoring of the amount and proportion of harmful factors, including heavy metals. The high risk of soil pollution with Cd, Zn, Ni and Pb as a result of industrial waste application as fertilizers was also evidenced [50].

Airborne sources of heavy metals include stack emissions or fugitive emissions such as dust from storage areas or waste heaps. Stack emissions can be distributed over a wide area by natural air currents, while fugitive emissions are often distributed over much smaller areas. In general, concentrations of pollutants are much lower in fugitive emissions compared to stack emissions. The type and concentration of metals emitted from both types of sources depend on site-specific conditions. All solid particles in smoke from fires and other emissions from factory chimneys are deposited on land or sea. Most forms of fossil fuels contain some heavy metals and this form of environmental pollution has been increasing since the industrial revolution began. For instance, very high concentrations of Cd, Pb and Zn have been found in plants and soils adjacent to smelting plants. Another major source of soil pollution is the aerial emission of lead from combustion of petrol containing tetraethyl lead; this contributes substantially to the content of Pb in soils in urban areas and in those adjacent to major roads [52].

Another, and one of the most significant sources of heavy metal pollution of soils, includes heavy industry, e.g. mining and metallurgy. Industrial airborne heavy metal contamination of the nonferrous smelters surrounding landscapes is a well-known and widely occurring phenomenon. Emissions of metallurgical dust are spread according to the wind direction and particle size while soil is the main receiver of heavy metals in dry land. Dust emissions from smelters using sulfide copper-nickel ores are similar, regardless of their location, owing to the fact that the same raw materials are used in metallurgical processes. The following major metal-containing compounds are deposited onto the landscape in the form of dust emissions from smelters: pentlandite $(Ni,Fe)9S8$, pyrrotite $Fe7Sg(Nix)$, chalcopyrite $CuFeS2$, chalcosite $Cu2S$, covellite CuS, cuprite $Cu2O$, tenorite CuO, and metal copper and nickel [54]. Surface soil layers in the mining or metallurgy areas are often heavily polluted with copper. In the vicinities of steel plants the concentration of this element exceeds several thousand ppm and the pollution remains for a long time, even after the operation of mines or steel plants had been stopped [50]. The fine fractions of dust are enriched with lead, arsenic, and zinc. The quantity and composition of dust derived from different sources (metallurgical processes) varies according to the raw materials and the condition of the gas cleaning systems [54]. The cause for the frequently widely dispersed metal pollution in habitats of mining areas was found in the formation of acid mine drainage (AMD). The runoff from mining heaps of active and abandoned mines can be extremely acidic, with pH values reaching as low as pH 2 [9]. Chemical and biological oxidation of the abundant mineral pyrite (FeS_2) occurs after the unearthing of pyrite-containing rock formations and results in an acidification of the dump material [55]. Under acidic conditions, the majority of heavy metals is leached from the waste dump and they are transported as AMD in streamwaters [9]. Galvanization industry may cause soil pollution with silver as well as other industrial facilities that use silver salts. Additionally, the increased amount of silver may by introduced to soils with municipal sewage. Municipal sewage contains also large amounts of highly soluble forms of zinc, which may then easily contaminate soil environment [50]. Zinc is also extensively used in metallurgical industry, as an anti-corrosion agent in alloys and in galvanization. It is frequently used in paint industry [50]. The concentration of cadmium highly increases in soils polluted with emissions from nonferrous metal plants, which constitute over 60% of all anthropogenic sources of this element in soils. Municipal sewage contains on average 10 – 40 ppm of cadmium, while industrial sewage may contain over 1000 ppm. This is also a case of large amounts of lead that may be introduced into soils from municipal sewage and waste, as they contain mobile forms of this element. This may result in large increase in the concentration of lead in soils that may exceed several times the admissible limits. Additionally, dust emissions from landfills of nonferrous metal plants may become dangerous sources of lead in soils [50]. Table 2 shortly summarizes the major sources of different heavy metals in soil.

Table 2: Different sources of heavy metals in soils [56]

Heavy metals	Sources
As	Semiconductors, petroleum refining, wood preservatives, animal feed additives, coal power plants, herbicides, volcanoes, mining and smelting
Cu	Electroplating industry, smelting and refining, mining, biosolids
Cd	Geogenic sources, anthropogenic activities, metal smelting and refining, fossil fuel burning, application of phosphate fertilizers, sewage sludge
Cr	Electroplating industry, sludge, solid waste, tanneries

Pb	Mining and smelting of metalliferous ores, burning of leaded gasoline, municipal sewage, industrial wastes enriched in Pb, paints
Hg	Volcano eruptions, forest fire, emissions from industries producing caustic soda, coal, peat and wood burning
Se	Coal mining, oil refining, combustion of fossil fuels, glass manufacturing industry, chemical synthesis (e.g., varnish, pigment formulation)
Ni	Volcanic eruptions, land fill, forest fire, bubble bursting and gas exchange in ocean, weathering of soils and geological materials
Zn	Electroplating industry, smelting and refining, mining, biosolids

THE EFFECTS OF HEAVY METALS ON SOIL MICROORGANISMS

Metals without biological function are generally tolerated only in minute concentrations, whereas essential metals with biological functions, are usually tolerated in higher concentrations [9]. They have either metabolic functions as constituents of enzymes or meet structural demands, e.g. by supporting the cell envelope. Frequently the concentration and the speciation of metal determine whether it is useful or harmful to microbial cells [9].

Microorganisms are the first biota that undergoes direct and indirect impacts of heavy metals. Some metals (e.g. Fe, Zn, Cu, Ni, Co) are of vital importance for many microbial activities when occur at low concentrations. These metals are often involved in the metabolism and redox processes. Metals facilitate secondary metabolism in bacteria, actinomycetes and fungi [9; 57]. E.g. chromium is known to have stimulatory effect on both actinorhodin production and growth yield of the model actinomycete S. coelicolor [58]. However, high concentrations of heavy metals may have inhibitory or even toxic effects on living organisms [59]. Adverse effects of metals on soil microbes result in decreased decomposition of organic matter, reduced soil respiration, decreased diversity and declined activity of several soil enzymes [60]. Some of the general changes in morphology, the disruption of the life cycle and the increase or decrease of pigmentation are easy to observe and evaluate [9]. Rajapaksha et al. [61] compared the reactions of bacteria and fungi to toxic metals in soils (Zn and Cu). They concluded, that bacterial community is more sensitive to increased concentrations of heavy metals in soils than the fungal community. The relative fungal/bacterial ratio increased with increasing metal levels. Those authors also noticed the varying effect of soil pH on the microbial reaction to soil pollution, i.e. that lower pH in contaminated soils enhanced the negative effect on bacteria, but not on fungi.

The toxic concentration of heavy metals may cause enzyme damage and consequently their inactivation, as the enzymes-associated metals can be displaced by toxic metals with similar structure [59]. Moreover, heavy metals alter the conformational structures of nucleic acids and proteins, and consequently form complexes with protein molecules which render them inactive. Those effects result in disruption of microbial cell membrane integrity or destruction of entire cell [62]. Heavy metals also form precipitates or chelates with essential metabolites [63].

Various metals may affect different microbial populations and the resulting impact may vary depending on the metal whose limit concentrations in soils were exceeded. For instance, the pollution of soils with copper affects microorganisms that take part in nitrification and mineralization of protein compounds [50]. Silver is one of the most toxic metals to heterotrophic bacteria. This effect is used for the production of antiseptic preparations. However, there are some silver-resistant bacteria, both in clinical and natural conditions. Some strains of *Thiobacillus ferrooxidans* are able to accumulate particularly large amounts of silver [50]. About 100 ppm of zinc in soils may inhibit nitrification processes and about 1000 ppm inhibits the majority of microbiological processes in soils [64]. Microorganisms play vital role in circulation and transformation of mercury compounds in the environment. Numerous bacteria and fungi show high tolerance (also acquired) to increased concentrations of mercury in soils. However, some microorganisms are sensitive to excess mercury, e.g. the concentration of <10 ppm Hg may have toxic effects on nitrifiers in soils [50]. Increased concentrations of lead in surface soil layers negatively affect soil microflora. Processes of organic matter decomposition, particularly cellulose, are inhibited as a result of decreased enzymatic activity of microorganisms. This results in soil degradation. Biosorption of lead by soil microorganisms reaches on average 0.2% of this metal, but in some cases it may reach even 40% of biomass and may be used for biological remediation [50]. Some studies indicate that long-term contamination of soils with heavy metals has adverse effects on soil microbial activity. For instance, Juwarkar et al. [65] while researching the remediation strategies for cadmium and lead contaminated soils, compared the numbers of the selected groups of microorganisms in natural and heavy metal spiked soils. The results that they obtained indicated that the examined microbial groups were much less abundant in contaminated soils than in natural ones [Table 3]. On the other hand, Lenart and WolnyKoładka [66] recorded significantly variable numbers of the selected microbial groups while analyzing the uncontaminated and heavy metal contaminated soils of ArcelorMittal steelworks in Cracow. Similarly, their results indicated that except for fungi, the soil-dwelling microorganisms were much less abundant in heavy metal polluted soils than in uncontaminated soils (Table 4). Heavy metal contamination results in reduction of microbial biomass and even if they do not cause the reduction in their number – they reduce biodiversity or disturb the community structure [64].

Table 3: Microbiological characteristics of natural and heavy metal spiked spoil samples in Nagpur (India) [65]

Analyzed microorganisms [CFU×g^{-1}]	Natural soil	Heavy metal contaminated soil
Total number of bacteria	17×10^5	58×10^4
Total number of fungi	26×10^3	42×10^2
Actinomycetes	43×10^3	18×10^1
Azotobacter spp.	23×10^3	17×10^1
Rhizobium spp.	21×10^3	16×10^2

Table 4: Ranges of the selected microbial groups in heavy metal contaminated and uncontaminated soils of ArcelorMittal steelworks in Cracow, Poland [66]

Analyzed microorganisms [CFU×g^{-1}]	Uncontaminated soil	Heavy metal contaminated soil
Total number of mesophilic bacteria	$22.50 \times 10^2 - 10.44 \times 10^6$	$0 - 13.15 \times 10^5$
Total number of fungi	$84.00 \times 10^1 - 21.03 \times 10^3$	$0 - 57.90 \times 10^3$
Actinomycetes	$62 - 99.50 \times 10^3$	$0 - 20.26 \times 10^3$
Azotobacter spp.	$0 - 28.90 \times 10^2$	$0 - 57.00 \times 10^1$

However, one of the reasons of decreasing biodiversity of microorganisms in heavy metal polluted soils is the selection for tolerant species or strains. Metal exposure may lead to the establishment of tolerant microbial populations, that are often represented by several Gram-positive genera such as *Bacillus*, *Arthrobacter* and *Corynebacterium* or Gram-negatives, e.g. *Pseudomonas*, *Alcaligenes*, *Ralstonia* or *Burkholderia* [68]. It was shown that the impact of heavy metals on the bacterial metabolism depends on the growth form. The resistance towards metals seems higher in consortia than in pure cultures [69]. A great number of heavy metal-resistant bacteria, such as e.g. *Cupriavidus metallidurans* possess efflux transporters that excrete toxic or overconcentrated metals outside the cell [70]. Efflux transporters have high substrate affinity and can therefore maintain low cytosolic concentration of metals [9]. Alternatively, microbial cells may prevent the intoxication by the release of metal-binding compounds into the extracellular surroundings. In that case, metals are chelated outside the cell and thus blocked from entering the cell through the membrane transporters that otherwise facilitate the influx [9]. Some fungal and bacterial species are able to keep metals outside their cells by the extracellularly active melanin [71]. It is a secondary metabolite that has strong cation chelating properties through the anionic function such as carboxyl and deprotonated hydroxyl groups [9]. A substantial number of soil microorganisms, such as widespread fungus *Aspergillus niger*, solubilize metals by the release of organic acids or by the immobilization of metals through excretion of different compounds, such as oxalates [72]. Some microorganisms possess the abilities to protect their cells by a cytosolic sequestration mechanisms. These mechanisms are activated once the metal enters the cell and cannot be excreted. In this case internal inclusion bodies, e.g. polyphosphate granules (volutin) bind large amounts of metal cations [73]. Investigation and understanding of microbial resistance mechanisms towards heavy metals are crucial for the potential application of microorganisms for remediation of polluted soils.

GENERAL OUTLINE OF SOIL REMEDIATION STRATEGIES

The overall objective of any soil remediation approach is to create a final solution that is protective both for human health and the environment [74]. For heavy metal-polluted soils, the physical and chemical form of the heavy metal contaminant in soil strongly influences the selection of the appropriate remediation treatment approach. Details on the physical characteristics of polluted soils, type and level of the pollution at the site must be known to enable accurate assessment of the problem severity and adjustment of remedial measures [52].

Remediation of heavy metal-polluted sites is very expensive and difficult, therefore the best method to protect the environment from contamination is to prevent it. Nevertheless, it is not always possible and once metals are introduced and pollute the soil, they will remain there. Unlike carbon-based organic pollutants, heavy metals cannot be degraded or eliminated completely, therefore the traditional treatments for heavy metal pollution of soils are complicated and cost-intensive.

There are several technologies for remediation of heavy metal-polluted soils. One of the classifications divides the methods into *in situ* and *ex situ* treatment technologies. *In situ* (in place) means that the polluted soil is treated in its original location, i.e. it remains at the site or in the subsurface. Such technologies remove the pollutant from soil without excavation or removal of the soil. In this case fixing agents are applied on the unexcavated soil. This technique's advantages may include low invasiveness, simplicity and rapidity. Moreover, it is fairly inexpensive and generates relatively low amount of waste. However, it is only a temporary solution. This is due to the fact that when physicochemical properties of soil change, the pollutants may again become active. Moreover, the reclamation process is applied only to the surface layer of soil [75]. *Ex situ* means that the treated soil is removed or excavated from the site [52]. It is applied in areas where heavily polluted soil must be removed from its place of origin and its storage is associated with high ecological risk. Fast and easy applicability, relatively low costs of investment and operation are the advantages of this method. On the other hand, it is highly invasive to the environment, generates a significant amount of solid wastes, and it is necessary to control the stored waste permanently. Evanko and Dzombak [76] divide *in situ*remediation strategies into solidification/stabilization, vitrification, soil flushing, electrokinetic extraction and biological treatment. *Ex situ* treatment technologies are divided by these authors into: solidification/stabilization, soil washing, vitrification and pyrometallurgical separation. Another classification of remedial strategies divides the technologies under five categories of general approaches to remediation: isolation, immobilization, toxicity reduction, physical separation, and extraction. There are several physicochemical techniques that include excavation and burial of soil at a hazardous waste site, chemical processing of soil to immobilize metals, leaching by using acid solutions or appropriate leachants to desorb metals from soil followed by the return of clean soil to the site [77], precipitation or flocculation followed by sedimentation, ion exchange, reverse osmosis and microfiltration [78]. Nevertheless, physicochemical techniques for heavy metal remediation are generally costly and have side effects [37]. Therefore, continuous efforts have been made to develop techniques that are easy to use, sustainable and economically feasible.

THE USE OF PLANTS FOR BIOLOGICAL REMEDIATION OF HEAVY METAL POLLUTED SOILS

Phytoremediation is one of the best techniques for treatment of heavy metal-polluted sites. It is an *in situ* strategy that uses vegetation and associated microbiota together with agronomic practices to aid in metal remediation [79]. It is based on the use of special type of plants to decontaminate soil by inactivating metals in the rhizosphere or translocating them in the aerial parts [56]. Some plants developed mechanisms to remove ions selectively from the soil to regulate the uptake and distribution of metals. Potentially useful phytoremediation technologies for heavy metal-polluted sites include phytoextraction, phytostabilization and rhizofiltration [75].

Phytoextraction uses hyperaccumulating plants to remove metals from soil by absorption into the roots and shoots of the plant. The aboveground shoots can be then harvested to remove metals from the site and subsequently stored as hazardous waste or employed for the recovery of metals. The ideal plant for phytoextraction should grow rapidly, produce high amount of biomass and be able to tolerate and accumulate high metal concentrations in shoots [80]. Hyperaccumulating plants belong to the families of *Brassicaceae, Fabaceae, Euphorbiaceae, Asterraceae, Lamiaceae,* and *Scrophulariaceae* [77]. Studies indicate that many *Brassica* species, such as *B. juncea, B. napus* or *B. rapa* exhibit enhanced accumulation of Zn and Cd [81] In comparison to conventional methods like e.g. soil excavation (*ex situ* remediation), phytoextraction is time consuming, but on the other hand it is cost-effective and less labor-intensive [9].

Phytostabilization is based on the use of plants to limit the mobility and bioavailability of metals in soil. Plants used in this method are characterized by high tolerance of metals in surrounding soils together with their low accumulation. Phytostabilization can be carried out through the process of sorption, precipitation, complexation, or metal valence reduction. This technique is useful for the removal of Pb, As, Cd, Cr, Cu, and Zn [82]. This process is advantageous because in this case disposal of hazardous material/biomass is not required, and it is very effective when rapid immobilization is needed to preserve soils or ground and surface waters [82].

Rhizofiltration (or phytofiltration) removes metals from contaminated soil via absorption, concentration and precipitation by plant roots. This technique is used to remove pollutants from groundwater and aqueous-waste streams rather than for the remediation of polluted soils [76]. Apart from the above described phytoremediation methods, some authors [83] include also phytovolatization and phytodegradation.

Phytovolatization involves the use of plants to volatilize pollutants from their foliage such as Se and Hg, while phytodegradation uses plants and associated microorganisms to degrade organic pollutants.

Even though phytoremediation strategies are inexpensive, effective, environmentally friendly and can be implemented *in situ*, a substantial proportion of metal pollutants are unavailable for root uptake by field grown plants [84]. Therefore, methods of increasing phytoavailability of heavy metal pollutants in soil and their transport to plant roots are vital to the success of *in situ* phytoremediation. In this case it is useful to apply microbial populations that are able to affect trace metal mobility and availability to plants, through the release of chelators, acidification and redox changes [85]. It was proved that the presence of rhizosphere bacteria increases the available concentrations of various heavy metals to hyperaccumulative plants [80]. Microbial populations may be used not only for increasing metal bioavailability to plants, but also for the promotion of hyperaccumulative plant growth through N_2 fixation, production of phytohormones and siderophores, and transformation of nutrients [26]. Figure 2 summarizes the mechanisms of plant-mediated remediation of contaminated soils.

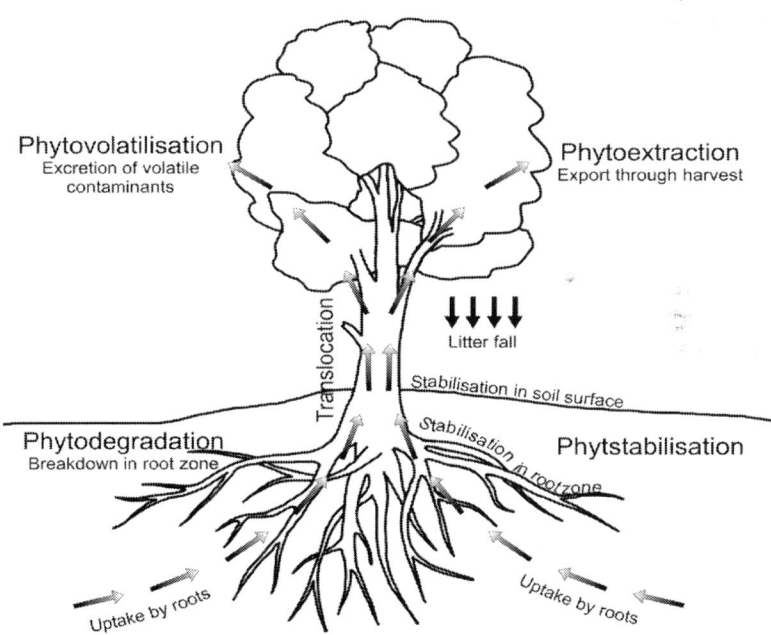

Figure 2: Mechanisms of phytoremediation involved in purifying contaminated soils and physiological processes that occur in plants during phytoremediation.

APPLICATION OF MICROORGANISMS TO REMEDIATE HEAVY METAL-POLLUTED SOILS

Another approach for biological remediation of heavy metal-polluted soils includes the use of microorganisms to detoxify metals by valence transformation, extracellular chemical precipitation or volatilization etc. [56]. Bioleaching is the method that uses microorganisms to solubilize heavy metal pollutants either by direct bacterial processes, or as a result of interactions with metabolic products, or both [76]. It can be used *in situ* or *ex situ* to help to remove the pollutants from soils. This process is based on mobilization of metal cations from insoluble ores by biological oxidation and complexation. This process was adapted from mining industry for the use in soil remediation and a general term covering both bioleaching and biooxidation techniques could be "biomining". This technique is mainly employed for copper, cobalt, nickel, zinc and uranium, which are extracted either from insoluble sulfides or (in the case of uranium) from oxides [86]. The classical bioleaching bacteria belong to the genus *Acidithiobacillus* (*A. thiooxidans* and *A. ferrooxidans*), *Acidiphilium*, *Acidimicrobium*, *Ferromicrobium* or *Sulfobacillus* [86].

Another solution for soil bioremediation using microorganisms is to apply microbially-mediated biochemical processes, such as oxidation/reduction or methylation reactions [87]. Often, biostimulation and bioaugmentation are the components of bioremediation strategies. Biostimulation is a form of *in situ* bioremediation which uses growth rate stimulation nutrients, electron donors or acceptors to encourage the growth of site-specific indigenous microorganisms capable of degrading environmental pollutants. Common electron donors and acceptors used in biostimulation include: acetate, sulfate, nitrate and ethanol [88]. Bioaugmentation is the introduction of specific competent microorganisms to the local microbial population in order to increase the metabolic capacities needed for remediation [89]. Biosorption is a physicochemical process that occurs naturally and allows to passively concentrate and bind contaminants onto the microbial cell structure [90]. Metal biosorption by living organisms is a complicated process that consists of two steps. In the first step, metal ions are adsorbed on the cell surface by interactions between metals and cell surface functional groups. Biosorption of metal ions occurs primarily on the outer surface of microbial cells and is the first step in the interactions between metals and microbial cell walls [4]. The cell wall consists of a variety of polysaccharides and proteins, and hence offers a number of active sites capable of binding metal ions [91]. Differences in the cell wall composition among various microbial groups, i.e. algae, bacteria, cyanobacteria and fungi, cause significant differences in the type and amount of metal ions binding to them [91]. Physical adsorption via electrostatic or van der Waals forces allow to retain metal ions on the outer surfaces of bacterial cells. In addition to physical adsorption, ion exchange and complexation are believed to be the dominant mechanisms involved in metal biosorption [4]. The first step, passive biosorption, is metabolism-independent and proceeds rapidly by any one or a combination of metal binding mechanisms. In the second step, due to active biosorption, metal ions penetrate the cell membrane and enter into the cells. This is, however, a slowly occurring process. Active mode is metabolism-dependent and related to metal transport and deposition [91]. There are several microbial genera and species capable of metal biosorption. Fungi were found to be efficient biosorbent organisms, as their cells are characterized by a high percentage of cell wall material, which shows excellent metal binding properties [92]. *Aureobasidium pullulans*, *Cladosporium resinae*, *Aspergillus niger*, *Aspergillus versicolor* or *Rhizopus nigricans* are the fungal species proved to be effective in heavy metal biosorption [91]. Numerous studies also identified several species of bacteria as efficient metal accumulating microorganisms. For instance, *Bacillus* spp. has been reported to have a high potential of metal sequestration and has been used in commercial biosorbent preparation [91]. Other bacterial species capable of metal transformation include, among others: *Escherichia coli*, *Pseudomonas maltophilia*, *Shewanella putrefaciens*, *Pseudomonas aeruginosa*, *Enterobacter cloacae* [4].

Mechanisms involved in biochemical interactions between bacteria and metal ions involve specific enzymes that catalyze the oxidation, reduction, methylation, dealkylation and precipitation reactions. Microorganisms transform a substantial number of metals and metalloids by reducing or oxidizing them directly to a lower or higher redox state. Additionally, indirect oxidation or reduction is an alternative for immobilization of toxic metals in the environment.

Methylation is an important process involved in geochemical cycling of metals and the removal of metal pollutants from soils. Methylation processes derive the methyl group from methylocarbolamine (CH_3B_{12}) which is implicated in the methylation of multiple metals and metalloids, such as Pb, Sn, Pd, Pt, Au, Ti, As, Se and Te [93]. Methylation of Hg, Sn and Pb can be mediated by a range of microbes, including *Clostridium* spp., methanogens and sulfate-reducing bacteria under anaerobic conditions and principally by fungi (e.g.*Penicillium* spp. and *Alternaria* spp.) under aerobic conditions. Methyl groups are enzymatically transferred to metals and a given species may transform a number of different metals [94]. Methyl-metal compounds are generally highly volatile and available to plants [50]. Another mechanism that has the potential for the application in heavy metal-polluted sites is the production of siderophores by different microbial genera. Siderophores are the largest class of compounds that can bind and transport Fe. They are highly specific Fe(III) ligands and are excreted by a wide variety of fungi and bacteria to aid Fe assimilation [94].

Microorganisms play an important role in the environmental biogeochemical cycle of metals and their properties are of significant interest in the remediation of contaminated sites. The microbial ability to absorb and transform metals is a promising aspect in respect of solving the pollution problems [4]. The potential of numerous microbial metal transformations in treatment of environmental pollution may be employed and some processes are already in commercial operation. However, many processes are still at the laboratory scale and yet to be tested in a rigorous applied and/or commercial context [94]. Another interesting aspect of the microbial community is their ability to multiply even under undesirable environmental conditions. These microorganisms sometimes affect soil environment more quickly than abiotic processes can. Therefore, the structure of soil microbial populations may be useful as a highly sensitive bioindicator of soil disturbance and progress of remediation [95].

Facing the increasing heavy metal pollution severity accompanied by rising land prices the communities around the world need to struggle for available investment grounds. This is mostly the problem of big cities, especially those with limited opportunities for development due to geographical barriers such as seashores, mountain ranges or desert areas. In such situations the polluted industrial areas cannot be left unused for long time to recover naturally. This creates a need for the development of various remedial procedures adjusted to changing contamination level, environmental conditions, available time and funding. Thus, remedial measures need to be almost always modified in order to meet those criteria. This makes that the continuous effort should be made to increase the effectiveness, flexibility and decrease the cost and side effects of the procedures available today. Although a number of measures was developed to remove the even toxic level of contamination, there are many degenerated areas that still cannot be successfully treated now. Those cases involve sites where remediation would be too expensive, time consuming or even technically disputable with currently available treatment procedures.

CONCLUSIONS

Heavy metals pose a significant threat towards the soil environment and the rapid industrialization will result in increasing problems of environmental pollution. Therefore, it is necessary to carry out the continuous monitoring of both industrial areas and their vicinities for possible transgressions of the limits given by the authorities. When necessary, the remedial measures should be applied as soon as possible by all available means. On the other hand, research should be promoted to understand the mechanisms of microbial response to heavy metal pollution and to enable screening for possible resistant microorganisms that could be used for both remediation and restoration of soil environment fertility.

REFERENCES

1.   Ehrlich H.L. Geomicrobiology. 4th ed. New York: Marcel Dekker; 2002.

2.   Appenroth K.J. Definition of "heavy metals" and their role in biological systems. In: Sherameti I., Varma A. (eds.) Soil heavy Metals, Soil Biology. Berlin: Springer; 2010. P. 19: 19-29.

3.   Szyczewski P., Siepak J., Niedzielski P., Sobczyński T. Research on heavy metals in Poland. Polish Journal of Environmental Studies 2009; 18: 755-768.

4.   Han X., Gu J.D. Sorption and transformation of toxic metals by microorganisms. In: Mitchell R., Gu J.D. (eds.) Environmental Microbiology 2nd ed. New Jersey: Wiley-Blackwell; 2010. p. 153–175.

5.   Tate R.L. Soil Microbiology. 2nd ed. Hoboken: John Wiley and Sons Inc. USA; 2000.

6.   McKinney R. E. Environmental Pollution Control Microbiology. New York: Marcel Dekker, Inc.; 2004.

7.   Roselló-Mora R., Amann R. The species concept for prokaryotes. FEMS Microbiology Reviews 2001; 25: 39-67.

8.   Metting F.B. Soil Microbial Ecology. New York: Marcel Dekker Inc.; 1993.

9.   Hafeburg G., Kothe E. Microbes and metals: interactions in the environment. Journal of Basic Microbiology 2007; 47:453-467.

10.   Ranjard L., Richaume A. Quantitative and qualitative microscale distribution of bacteria in soil. Research in Microbiology 2001; 152: 707-716.

11.   Torsvik V., Øvreås L. Microbial diversity and function in soil: from genes to ecosystems. Current Opinion in Microbiology 2002; 5:240-245.

12.   Sessitsch A., Weilhalter A., Gerzabek M.H., Kirchmann H., Kandeler E. Microbial population structures in soil particle size fractions of a long-term fertilizer field experiment. Applied Environmental Microbiology 2001; 67: 4215-4224.

13.   Grayston S.J., Griffith G.S., Mawdsley J.L., Campbell C.D., Bardgett R.D. Accounting for variability in soil microbial communities of temperate upland grassland ecosystems. Soil Biology and Biochemistry 2001; 33: 533-551.

14.   Fierer N., Jackson R.B. The diversity and biogeography of soil bacterial communities. Proceedings of the National Academy of Sciences 2006; 103: 626-631.

15.   Hoorman J.J. The Role of Soil Bacteria. Fact Sheet – Agriculture and Natural Resources. The Ohio State University; 2011.

16.   Agriinfo. Soil Microorganism: Bacteria. http://agriinfo.in/default.aspx?page=topic&superid=5&topicid=147 (accessed 10 July 2013)

17. Jenkins A. Soil fungi. In: Soil biology basics. State of New South Wales, Department of Primary Industries; 2005. http://www.dpi.nsw.gov.au/__data/assets/pdf_file/0020/41645/Soil_fungi.pdf (accessed 20 June 2013).

18. Schlegel H.G. General microbiology. Cambridge: Cambridge University Press; 1993.

19. Agriinfo. Soil Microorganism: Algae. http://agriinfo.in/?page=topic&superid=5&topicid=150 (accessed 10 July 2013).

20. Barabasz W., Albi ska D., Ja kowska M., Lipiec J. Biological effects of mineral nitrogen fertilization on soil microorganisms. Polish Journal of Environmental Studies 2002; 11: 193-198.

21. Lynch J.M. The terrestrial environment. In: Lynch J.M., Hobbie J.E., (eds.) Microorganisms in Action: Concepts and Applications in Microbial Ecology. Oxford: Blackwell; 1988. p. 67-91.

22. Kloepper J.W., Leong J., Teintze M., Schroth M.N. Enhanced plant growth by siderophores produced by plant growth-promoting rhizobacteria. Nature 1980; 286: 885-886

23. Rodriguez H., Fraga R. Phosphate solubilizing bacteria and their role in plant growth promotion. Biotechnology Advances 1999; 17: 319-339.

24. Gholami A., Shahsavani S., Nezarat S. The effect of Plant Growth Promoting Rhizobacteria (PGPR) on germination, seedling growth and yield of maize. World Academy of Science, Engineering and Technology 2009; 25: 19-24.

25. Do Vale Barreto Figueiredo M., Seldin L., de Araujo F.F., de Lima Ramos Mariano R. Plant Growth Promoting Rhizobacteria: fundamentals and applications. In: Maheshwari D.K. (ed.) Plant Growth Promoting Bacteria, Microbiology Monographs 18, Berlin Heidelberg: Springer-Verlag; 2010. p. 21-43.

26. Glick B.R., Patten C.L., Holgin G., Penrose D.M. Biochemical and genetic mechanisms used by plant growth promoting bacteria. London: Imperial College Press; 1999.

27. Young J.M., Kuykendall L.D., Martínez-Romero E., Kerr A., Sawada H. A revision of *Rhizobium* Frank 1889, with an emended description of the genus, and the inclusion of all species of *Agrobacterium* Conn 1942 and *Allorhizobium undicola* de Lajudie et al. 1998 as new combinations:*Rhizobium*

radiobacter, R. rhizogenes, R. rubi, R. undicola and *R. vitis*. International Journal of Systematic and Evolutionary Microbiology 2001; 51:89-103.

28. Viss W.J., Pitrak J., Humann J., Cook M., Driver J., Ream W. Crown-gall-resistant transgenic apple trees that silence Agrobacterium tumefaciens oncogenes. Molecular Breeding 2003; 12: 283-295.

29. Toth I.K., Bell K.S., Holeva M.C., Birch P.R. Soft rot erwiniae: from genes to genomes. Molecular Plant Pathology 2003; 4: 17-30.

30. Koike S.T., Subbarao K.V., Davis R.M., Turini T.A. Vegetable diseases caused by soilborne pathogens. Publication 8099 of the regents of the University of California, Division of Agriculture and Natural Resources, US; 2003. http://anrcatalog.ucdavis.edu/pdf/8099.pdf (accessed 21 June 2013).

31. Williams R.E., Shaw III C.G., Wargo P.M., Sites W.H. Armillaria Root Disease. Forest Insect and Disease Leaflet 78. U.S. Department of Agriculture Forest Service; 1989. http://na.fs.fed.us/spfo/pubs/fidls/armillaria/armillaria.htm (accessed 16 July 2013).

32. Moss M.O., Smith J.E. The applied mycology of Fusarium. Cambridge: Cambridge University Press; 1984.

33. Vartivarian S.E., Anaissie E.J., Bodey G.P. Emerging fungal pathogens in immunocompromised patient: classification, diagnosis and management. Clinical Infection and Disease 17 (Suppl 2) 1993; S487-491.

34. Ryan J.R. Clostridium, Peptostreptococcus, Bacteroids, and other Anaerobes. In: Ryan K. J., Ray C. G. (eds.) Sherris Medical Microbiology: An Introduction to Infectious Diseases 4th ed.. Columbus: McGraw-Hill; 2004. pp. 309-326.

35. Maier R.M., Pepper I.L., Gerba C.P. Environmental Microbiology. Philadephia: Elsevier Inc.; 2009.

36. Pečiulyté D., Dirginčiuté-Volodkiené V. Effect of long-term industrial pollution on soil microorganisms in deciduous forests situated along a pollution gradient next to a fertilizer factory. Ekologija 2009; 55: 67-77.

37. McGrath S.P., Zhao F.J., Lombi E. Plant and rhizosphere process involved in phytoremediation of metal-contaminated soils. Plant and Soil 2001; 232: 207-214.

38. Gisbert C., Ros R., de Haro A., Walker D.J., Pilar Bernal M., Serrano R., Avino J.N. A plant genetically modified that accumulates Pb is especially promising for phytoremediation. Biochemical and Biophysical Research Communications 2003; 303: 440-445.

39. Bilos C., Colombo J.C., Skorupka C.N., Rodriguez Presa M.J.. Sources, distribution and variability of airborne trace metals in La Plata City area, Argentina. Environmental Pollution 2001; 111: 149-158.

40. Pierzy ski G.M., Sims J.T., Vance G.F. Soils and Environmental Quality. London: CRC Press; 2000.

41. D'Amore J.J., Al-Abed S.R., Scheckel K.G., Ryan J.A. Methods for speciation of metals in soils: a review. Journal of Environmental Quality 2005; 34: 1707-1745.

42. Kaasalainen M., Yli-Halla M. Use of sequential extraction to assess metal partitioning in soils. Environmental Pollution 2003; 126: 225-233.

43. Indeka L., Karczun Z. Accumulation of selected heavy metals in soils along busy traffic routes. Ecology and Technology 1999; 6: 174-180.

44. Antonkiewicz J., Macuda J. Levels of heavy metals and hydrocarbons in grounds surrounding some petrol stations in Kraków. Acta Scientiarum Polonorum, Formatio Circumiestus 2005; 4 (2): 31-36.

45. Voegborlo R.B., Chirgawi M.B. Heavy metals accumulation in roadside soil and vegetation along major highway in Libiya. Journal of Science and Technology 2007; 27: 1-12.

46. Arslan H., Gizir A.M. Heavy-metal content of roadside in Mesin, Turkey. Fresenius Environmental Bulletin 2006; 15: 15-20.

47. Atafar Z., Mesdaghinia A., Nouri J., Homaee M., Yunesian M., Ahmadimoghaddam M., Mahvi A.H. Effect of fertilizer application on soil heavy metal concentration. Environmental Monitoring Assessment 2010; 160: 83-89.

48. Basta N.T., Ryan J.A., Chaney R.L. Trace element chemistry in residual-treated soil: key concepts and metal bioavailability. Journal of Environmental Quality 2005; 34: 49-63.

49. Alloway B.J. Soil processes and the behavior of metals., New York: Wiley; 1995.

50. Kabata-Pendias A., Pendias H. Biogeochemistry of trace elements. Warsaw: PWN Scientific Publishing House; 1999.

51. Du Plessis K.R., Botha A., Joubert L., bester R., Conradie W.J., Wolfaardt G.M. Response of the microbial community to copper oxychloride in acidic sandy loam soil. Journal of Applied Microbiology 2005; 98: 901-909.

52. Wuana R.A., Okieimen F.E. Heavy metals in contaminated soils: a review of sources, chemistry, risks and best available strategies for remediation. ISRN Ecology 2011. doi:10.5402/2011/402647.

53. De Vries W.R., Römkens P.F.A.M., Van Leeuwen T., Bronswijk J.J.B. Heavy metals. In: Haygarth P.M., Jarvis S.C. (eds.) Agriculture, hydrology and water quality. Nosworthy Way: CABI ; 2002. p. 107-132.

54. Barcan V. Leaching of nickel and copper from soil contaminated by metallurgical dust. Environment International 2002; 28: 63-68.

55. Colmer A.R., Hinkel M.E. The role of microorganisms in acid mine drainage: a preliminary report. Science 1947; 106: 253-256.

56. Lone M.I., He Z., Stoffella P.J., Yang X.. Phytoremediation of heavy metal polluted soils and water: progresses and perspectives. Journal of Zhejiang University Science B (Biomedicine & Biotechnology) 2008: 9: 210-220

57. Weinberg E.D. Roles of trace metals in transcriptional control of microbial secondary metabolism. Biology of Metals 1990; 2: 191-196.

58. Abbas A.S., Edwards C. Effects of metals on Streptomyces coelicolor growth and actinorhodin production. Applied Environmental Microbiology 1990; 56: 675-680.

59. Bruins M.R., Kapil S., Oehme F.W. Microbial resistance to metals in the environment. Ecotoxicology and Environmental Safety 2000; 45: 198-207.

60. Tyler G. Heavy metal pollution and soil enzymatic activity. Plant and Soil 1974; 41: 303-311.

61. Rajapaksha R.M.C.P. Tabor-Kapłon M.A., Bååth E. Metal toxicity affects fungal and bacterial activities in soil differently. Applied and Environmental Microbiology 2004; 70: 2966-2973

62. Bong C.W., Malfatti f., Azam F., Obayashi Y., Suzuki S. The effect of zinc exposure on the bacteria abundance and proteolytic activity in seawater. In: Hamamura N., Suzuki S., Mendo S., Barroso C.M., Iwata H., Tanabe S. (eds.) Interdisciplinary Studies on Environmental Chemistry – Biological Responses to Contaminants. Tokyo: Terrapub; 2010. p. 57-63.

63. Sobolev D., Begonia M.F.T. Effects of heavy metal contamination upon soil microbes: lead-induced changes in general and denitrifying microbial communities as evidenced by molecular markers. International Journal of Environmental Research and Public Health 2008; 5: 450-456.

64. Doelman P. Resistance of soil microbial communities to heavy metals. In: Jensen V., Kioller A., Sorensen C.H. (eds.) Microbial communities in soil. London: Elsevier Applied Science Publishers; 1986. p. 369–384.

65. Juwarkar A.A., Nair A., Dubey K.V., Singh S.K., Devotta S. Biosurfactant technology for remediation of cadmium and lead contaminated soils. Chemosphere 2007; 68: 1996-2002

66. Lenart A., Wolny-Kołdka K. The effect of heavy metal concentration and soil pH on the abundance of selected microbial groups within ArcelorMittal Poland steelworks in Cracow. Bulletin of Environmental Contamination and Toxicology 2013; 90: 85-90

67. Wyszkowska J., Kucharski J., Borowik A., Boros E. Response of bacteria to soil contamination with heavy metals. Journal of Elementology 2008; 13: 443-453.

68. Piotrowska-Seget Z., Cyco M., Kozdrój J. Metal-tolerant bacteria occurring in heavily polluted soil and mine spoil. Applied Soil Ecology 2005; 28: 237-246.

69. Sprocati A.R., Alisi C., Segre L., Tasso F., Galletti M., Cremisini C. Investigating heavy metal resistance, bioaccumulation and metabolic profile of a metallophile microbial consortium native to abandoned mine. Science of the Total Environment 2006; 366; 649-658.

70. Nies D.H. Efflux-mediated heavy metal resistance in prokaryotes. FEMS Microbiology Reviews 2003; 27: 313-339.

71. Fogarty R.V, Tobin J.M. Fungal melanins and their interactions with metals. Enzyme and Microbial Technology 1996; 19: 311-317.

72. Gadd G.M. Fungal production of citric and oxalic acid: importance in metal speciation, physiology and biogeochemical processes. Advances in Microbial Physiology 1999; 41: 47-92.

73. Gonzalez H., Jensen T.E. Nickel sequestering by polyphosphate bodies in Staphylococcus aureus. Microbios 1998; 93: 179-185.

74. Martin T.A., Ruby M.V. Review of in situ remediation technologies for lead, zinc and cadmium in soil. Remediation 2004; 14: 35-53.

75. USEPA. Recent developments for in situ treatment of metal contaminated soils. Tech. Rep. EPA-542-R-97-004, Washington DC: USEPA; 1997. http://www.clu-in.org/download/remed/metals2.pdf (accessed 15 June 2013).

76. Evanko C.R., Dzombak D.A. Remediation of metals-contaminated soils and groundwater. Technology Evaluation Report. Pittsburgh: Ground-Water Remediation Technologies Analysis Center; 1997.

77. Salt D.E., Smith R.D., Raskin I. Phytoremediation. Annual Reviews in Plant Physiology & Plant Molecular Biology 1998; 49: 643-668.

78. Raskin I., Smith R.D., Salt D.E. Using plant seedlings to remove heavy metals from water. Plant Physiology 1997; 111: 552-552.

79. Cunningham S.D., Ow D.W. Promises and prospects of phytoremediation. Plant Physiology 1996; 110: 715-719.

80. Jing Y., He Z., Yang X. Role of soil rhizobacteria in phytoremediation of heavy metal contaminated soils. Journal of Zhejiang University Science B (Biomedicine & Biotechnology) 2007; 8: 192-207.

81. Ebbs S.D., Lasat M.M., Brady D.J., Cornish J., Gordon R., Kochian I.V. Phytoextraction of cadmium and zinc from a contaminated soil. Journal of Environmental Quality 1997; 26: 1424-1430.

82. Jadia C.D., Fulekar M.H. Phytoremediation of heavy metals: recent techniques. African Journal of Biotechnology 2009; 8: 921-928.

83. Garbisu C, Alkorta I. Phytoextraction: A cost effective plant-based technology for the removal of metals from the environment. Bioresource Technology 2001; 77: 229-236.

84. Ajaz Haja Mohideen R., Thirumalai Arasu V., Narayanan K.R., Zahir Hussain M.I. Bioremediation of heavy metal contaminated soil by the Exigobacterium and accumulation of Cd, Ni, Zn and

Cu from soil environment. International Jornal of Biological Technology 2010; 1: 94-101.

85. Smith S.E., Read D.J. Mycorrhizal symbiosis. San Diego: Academic Press Inc.; 1997.

86. Rohwerder T., Gehrke T., Kinzler K., Sand W. Bioleaching review part A: Progress in bioleaching: fundamentals and mechanisms of bacterial metal sulfide oxidation. Applied Microbiology and Biotechnology 2003; 63: 239-248.

87. Means J.L., Hinchee R.E. Emerging technology for bioremediation of metals. Boca Raton: Lewis Publishers; 1994.

88. Miller H. Biostimulation as a form of bioremediation of soil pollutans. Basic Biotechnology eJournal 2010 http://ejournal. vudat.msu.edu/index.php/mmg445/article/viewArticle/ MMG445.4543071/395 (accessed 10 July 2013).

89. Gentry T.J., Rensing C., Pepper I.L. New approaches for bioaugmentation as a remediation technology. Critical Reviews in Environmental Science and Technology 2004; 34: 447-494.

90. Volesky B. Biosorption of heavy metals. Florida: CRC Press; 1990.

91. Das N., Vimala R., Karthika P. Biosorption of heavy metals – an overview. Indian Journal of Biotechnology 2008; 7: 159-169.

92. Horikoshi T., Nakajima A., Sakaguchi T. Studies on the accumulation of heavy metal elements in biological systems: accumulation of uranium by microorganisms. European Journal of Applied Microbiology and Biotechnology 1981; 12: 90-96.

93. Ridley W.P. Dizikes L.J., Wood J.M. Biomethylation of toxic elements in environment. Science 1977; 197: 329-332.

94. Gadd G.M. Metals, minerals and microbes: geomicrobiology and bioremediation. Microbiology 2009; 156: 609-643.

95. Gremion F., Chatzinotas A., Kaufmann K., Sigler W.V., Harms H. Impacts of heavy metal contamination and phytoremediation on a microbial community during a twelve-month microcosm experiment. FEMS Microbiology Ecology 2004; 48: 273-283.

Chapter 7

Sol-gel Synthesized Semiconductor Oxides in Photocatalytic Degradation of Phenol

Maria K. Cherepivska and Roman V. Prihod'ko

Dumanskii Institute of Colloid and Water Chemistry, National Academy of Sciences of Ukraine, Kiev, Ukraine

ABSTRACT

Effectiveness of photocatalytic degradation of phenol in aqueous solution using semiconductor oxides (SO) prepared by a sol-gel method was examined. The physical and chemical properties of synthesized catalysts were investigated by X-ray diffraction (XRD), diffuse reflectance UV-Vis spectroscopy (DRS), and N_2-adsorption measurements. The optimal conditions of the photocatalytic degradation of phenol using prepared titanium dioxide sample were defined.

INTRODUCTION

Heterogeneous photocatalysis on the semiconductors allows achieving complete mineralization of the various classes toxic and biorefractory organic substances [1, 2]. Recently, the photocatalytic degradation of toxicants became one of the most promising directions of "green chemistry" [3]. The use of nanosized SO presents a great interest due to their outstanding optical, magnetic, catalytic, and sensing properties [4, 5]. The phenolic compounds contained in the wastewater of chemical, petrochemical, and pharmaceutical industries are hazardous carcinogenic and mutagenic pollutants [6, 7]. Furthermore, the oxidation of these substances in water bodies leads to decrease in dissolved oxygen required for normal functioning of animals and plants. Finding effective methods for the protection of water systems from phenols contamination is an important aim to ensure environmental safety [8, 9].

Among SO photocatalysts (PC) high activity have Fe_2O_3, WO_3, ZnO and TiO_2. Iron oxide polymorphs of hematite ($\alpha-Fe_2O_3$) are nontoxic, cheap, and stable to photocorrosion material intensively absorbs radiation in the range from 295 to 600 nm. The semiconductor properties of $\alpha-Fe_2O_3$ are the same as WO_3, which can be seen in the position of band gaps relative to the standard hydrogen electrode. WO_3 has chemical stability in acidic medium and electrolyte solutions as well as photoactivity in the near ultraviolet and blue regions of solar spectrum [10]. According to Daneshvar et al. nanosized ZnO is a suitable alternative to TiO_2 due to the band gap energy. Dinda and Icli found that ZnO was as reactive as TiO_2 for the photocatalytic degradation of phenol under concentrated sunlight [11]. Figure 1 shows a scheme of the energy levels of the studied semiconductor oxides relative to the standard hydrogen potential [12]. Several authors have associated the efficiency of semiconductor photocatalysts with electronic, structural, and morphological properties of the material such as band gap energy, crystalline structure, surface area, particle size [13].

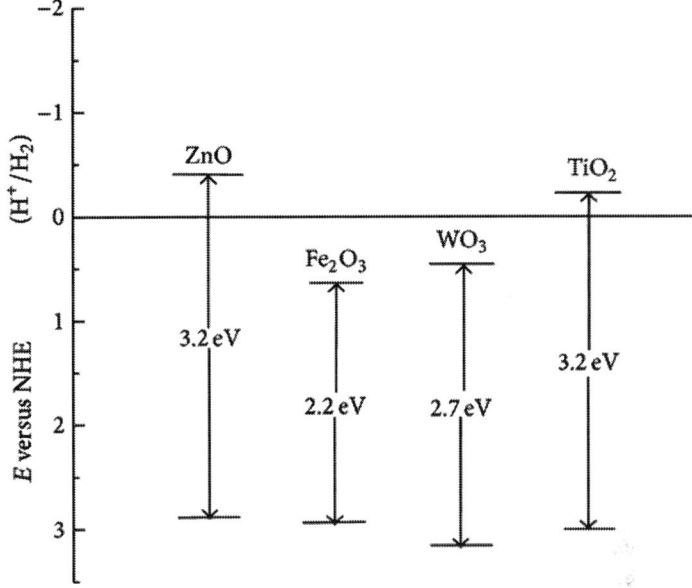

Figure 1: Energy band gap of investigated semiconductor oxides.

The activity of semiconductor oxides prepared by the sol-gel methods was investigated under the same conditions for searching of the most effective system in the reaction of the phenol photodegradation. The optimal parameters of the phenol photodegradation on the synthesized PC were defined.

MATERIALS AND METHODS

Materials

All solvents and chemicals used in this work were of analytical grade and were used without further purification. Inorganic (Fe(NO$_3$)$_3$·9H$_2$O, Zn(NO$_3$)$_2$, Na$_2$WO$_4$·2H$_2$O), and organic (Ti(i-OPr)$_4$) precursors for the synthesis of oxides were purchased from Sigma-Aldrich. The homogeneous precipitant urea (CO(NH$_2$)$_2$) and nitric acid (HNO$_3$) were purchased from Merck (Germany). In some cases the pH solutions were varied with NaOH and H$_2$SO$_4$ (Merck, Germany). P25 TiO$_2$ was purchased from the Degussa Company in Germany and was used as the reference sample. The Millipore Q Millipore system was used for water purification.

Synthesis of Semiconductor Oxides Powders

Synthesis of PC was carried out by sol-gel method which allows obtaining nanosized metal oxide particles with desired structural and morphological properties [14].

Fe_2O_3 sample was synthesized similar to the method [15]. Solutions of $Fe(NO_3)_3$ and $CO(NH_2)_2$ were slowly added in the heated deionized water with continues stirring. The mixture was heated at 363 K for 5 h to form and aging of $Fe(OH)_3$ sol. The resulting russet precipitation was dried at 383 K and calcined at 773 K for 2 h in the air.

WO_3 sample was prepared by thermal decomposition of tungstic acid obtained by the sol-gel method [13]. In this procedure, sodium tungstate dihydrate was dissolved in deionized water under continuous stirring at 353 K. After total dissolution, concentrated nitric acid was added dropwise to the sodium tungstate solution. The reaction mixture was subjected to the aging process during 40 minutes at 353 K and left for 24 h at 293 K. The resulting pale yellow precipitation was filtered, washed with deionized water, dried at 353 K, and calcined at 773 K during 2 h in the air.

Synthesis of nanocrystalline ZnO powder was performed similar to [16], by alkaline hydrolysis of zinc nitrate in thermoinitiated decomposition of urea. The aqueous solutions of $Zn(NO_3)_2$ and $CO(NH_2)_2$ was heated at a temperature of 363 K and stirred for 24 h. The resulting precipitation was washed, dried at 383 K, and calcined at 773 K for 2 h.

Nanocrystalline TiO_2 particles were synthesized by the hydrolysis of titanium isopropoxide [17] using a modified method [18]. Synthesis was carried out under vigorous stirring in excess of 2-propanol. The reaction mixture was gradually heated to 358 K with addition of deionized water to eliminate the intermediate gelation process. The resulting solid precipitation was filtered, dried in the air at 358 K, and calcined at 773 K.

As a reference sample used a titanium dioxide Degussa P25 (TiO_2 P25), obtained by high-temperature gas-phase oxidation of titanium tetrachloride vapors [19].

Photocatalyst Characterizations Techniques

In order to characterize the powders instrument measurements were performed with X-ray diffraction (diffractometer DRON 3 M generating CoK_{α_1} (λ=0.17902 nm) radiation), diffuse reflectance UV-Vis spectroscopy (Shimadzu UV-2405 spectrometer with integrated sphere ISR-2200 and $BaSO_4$ as the reference), and N_2-adsorption measurements (Micromeritics ASAP vacuum device 2010).

Photocatalytic Reactor and Experimental Procedure

Photocatalytic degradation of phenol was carried out in a 0.5 L quartz reactor with a jacket under air bubbling (velocity 50 mL·min⁻¹) and temperature from 293 K to 323 K. The reaction mixture was agitated with a magnetic stirrer (800 rpm·min⁻¹). The concentrations of phenol and catalyst were 0.532 mM and 1 g·L⁻¹, respectively. Low-pressure mercury lamp DRB-8 submerged in a quartz casing used as a UV-radiation source with a maximum emission output at 254 nm. Reaction time was 3 hours. The separation of reaction mixture was performed by centrifugation. The effectiveness of photocatalytic process was evaluated relative to the photolysis carried out under similar conditions without catalyst usage. The phenol conversion was determined by the aromatic content recorded by the absorbance of the solution at 270 nm (C_{270}) with a Shimadzu UV-2405 spectrometer and concentration of total organic carbon (C_{TOC}) measured by Shimadzu TOC-VCSN analyzer.

RESULTS AND DISCUSSION

Photocatalysts Characterization

Figure 2 presents the X-ray powder diffraction patterns of the synthesized semiconductor oxides and reference sample TiO_2 P25.

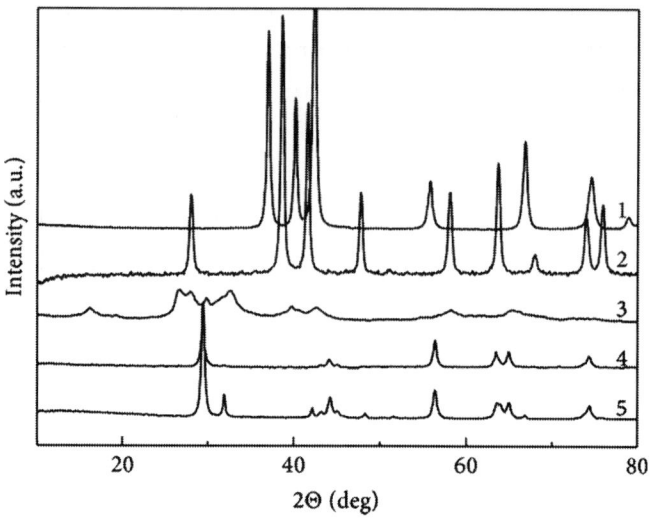

Figure 2: X-ray diffraction patterns of the semiconductor oxides: 1—ZnO; 2—α-Fe$_2$O$_3$; 3—WO$_3$; 4—TiO$_2$; 5—TiO$_2$ P25.

The XRD pattern of Fe$_2$O$_3$ sample shows that all basal reflections in the range of Bragg angles (2θ) from 10 to 80 characterize of isomorphic hematite phase (α-Fe$_2$O$_3$), corresponding to the orthorhombic crystal system (JCPDS No. 79-1741).

Investigation of the crystal structure of WO$_3$ confirms the presence of hexagonal phase (R6/mmm), JCPDS No. 33-1387). Low peak-height indicates the weakly crystallized structure.

The X-ray diffraction pattern of the prepared TiO$_2$ sample is presented basal reflections (at around 2θ 25.4, 44.2, and 56.4) corresponding to the titanium dioxide anatase phase [17].

As is known, after calcination for 2 h TiO$_2$ P25 is a mixture of anatase and rutile phase (82 and 18%, resp.) [19].

XRD analysis of the synthesized ZnO shows strong and high peaks indicating the high purity and crystallinity. Location of the basal reflections confirms hexagonal structure of ZnO (JCPDS No. 80-0075).

Diffuse reflectance spectroscopy allows obtaining information about light absorption range and band gap of the semiconductor [20]. Figure 3 depicts diffuse reflectance spectra of the synthesized materials converted in accordance to the Kubelka-Munk function. The energy band gap (E_g, eV) is determined by extrapolating of the onset of the rising part to x-axis (l_g, nm) of the plots by the dotted line and calculation by the following equation (see [21]):

$$E_g = \frac{1240}{\lambda_g}.$$

(1)

It can be clearly seen from Figure 3 that the radiation absorption by WO_3 sample begins in the visible range at 500 nm. The band gap of WO_3 is 2.55 eV which is consistent with literature data [22].

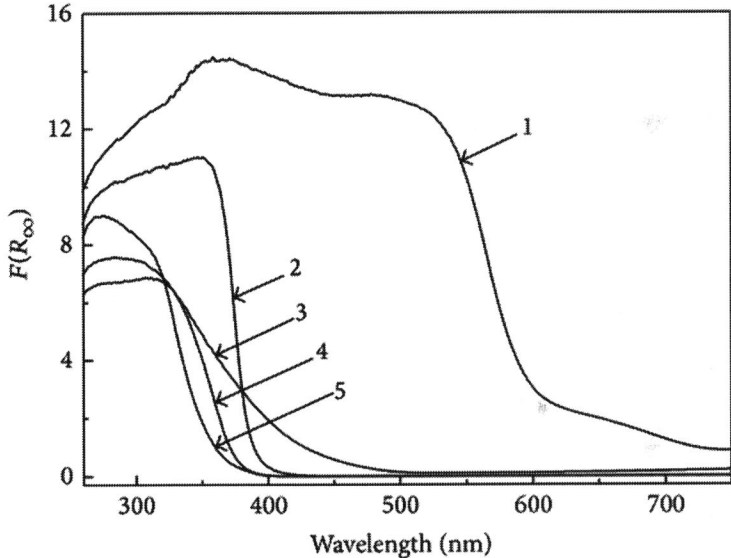

Figure 3: Diffuse reflectance spectra of semiconductor oxides: 1—α-F_2O; 2—ZnO; 3—WO_3; 4—TiO_2; 5—TiO_2 P25.

The optical absorption spectrum of the ZnO sample is represented by a broad and intense band and characterized by a sharp increase of absorption at 400 nm and a slight decrease at shorter wavelengths.

In the DRS of the TiO_2 P25 and synthesized TiO_2 powder the drastic increasing of the light absorption at λ=380 nm corresponding to the energy band gap of pure anatase (~3.2 eV) can be noted [23].

The surface area and micropore volume of the synthesized materials are defined with nitrogen adsorption-desorption isotherm. Surface parameters and the energy band gap of the SO are shown in Table 1. It is known that α-Fe_2O_3 and ZnO particles obtained by a sol-gel method by means of thermoinitiated decomposition of urea have a relatively large size (2000 and 3000 nm, resp.) [15, 16] and a low specific surface area. The synthesized WO_3 sample has a high specific surface area with a small volume share of the micropores. Despite the high dispersion of the synthesized TiO_2 sample, its specific surface area is less than that of the TiO_2 P25 sample because of lower specific volume of the micropores.

Table 1: Physical and chemical characteristics of the investigated semiconductor oxides

Sample	SBET (m2·g−1)	Vmicropore, (cm3·g−1)	Eg, (eV)	L a, (nm)
-Fe2O3	25,7	0,18	2,0	2000
WO3	50	0,005	2,55	20
ZnO	41	—d	3,23	3000
TiO2	45,3	0,1	3,43	5–7
TiO2 P25b	52	0,18	3,23	28

S_{BET}: specific surface area data obtained from the BET-model.

$V_{micropore}$: micropore volume data calculated by deBoer's t-plot method.

E_g: energy band gap.

L: linear particle size.

—: not determined.

[a,b]Published data.

Photodegradation of Phenol

The results of photolysis and photocatalytic degradation study of phenol using prepared semiconductor oxides and TiO_2 P25 samples are shown in Figure 4.

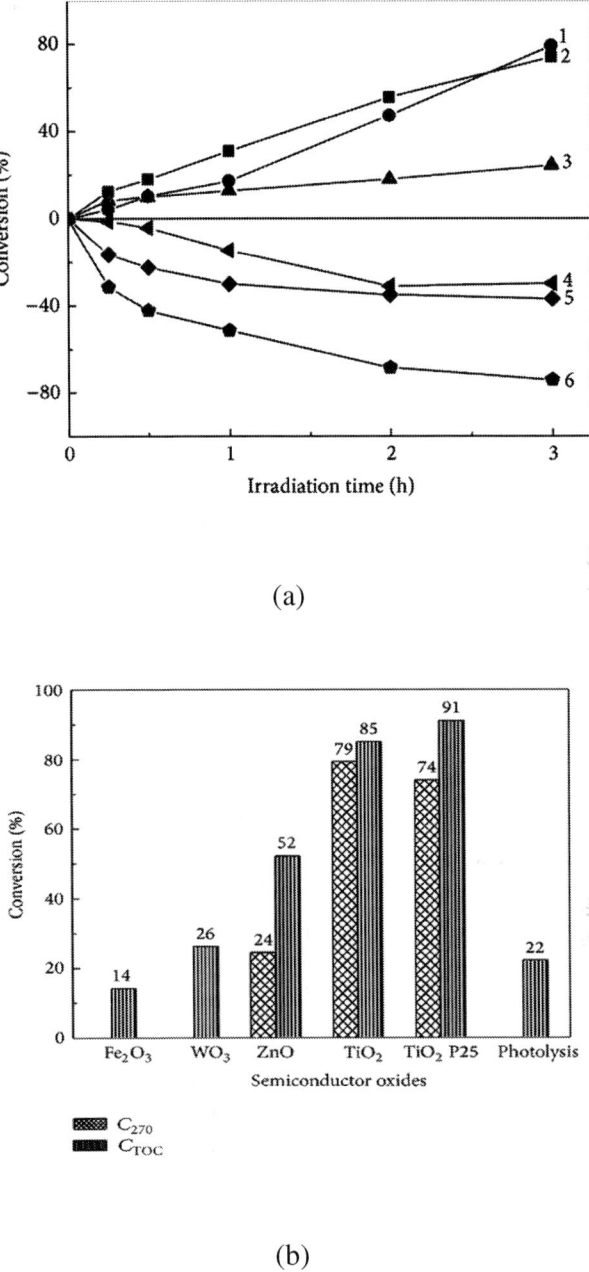

(a)

(b)

Figure 4: Effectiveness of investigated semiconductor oxides in photocatalytic phenol degradation. Experimental conditions: initial phenol

concentration = 0.536 mM; catalyst concentration = 0.1 g·L⁻¹; pH = 5.9; K. (a) Variations of C_{270} values during phenol photodegradation: 1—TiO₂ P25; 2—TiO₂; 3—ZnO; 4—WO₃; 5—photolysis; 6—α-Fe₂O₃; (b) C_{270} and C_{TOC} values of phenol solution after reaction.

During the photolysis of phenol under UV-C irradiation the appearance of light brown color and increase of the optical density of analyzed solution are observed, which can be explained by the formation of colored intermediates: benzoquinone, hydroquinone, and catechol [6]. Incomplete oxidation of phenol confirmed its low mineralization (22%, Figure 4), and indicates necessity of catalytic method usage.

The study of phenol conversion dependence on the SO nature found that the least active are the α-Fe₂O₃ and WO₃ samples (mineralization is 14% and 26% resp.). The increase of optical density of the solution after photocatalysis suggests the formation of colored intermediates. The phenol conversion using ZnO powder was 24% of aromatic content and 52% of TOC. TiO₂ samples showed the highest activity. Application of synthesized sample TiO₂ leads to aromatic content of 79% and TOC of 85% removal. These results are similar to the activity of TiO₂ P25 sample.

Table 2 shows the phenol aromatic content (C_{270}) and TOC (C_{TOC}) conversion dependence on the SO nature, as well as the total (A) and specific (a) catalytic activity of investigated SO, calculated by the following equation (see [24]):

$$A = \frac{C_{TOC}}{\tau \cdot m},$$

$$a = \frac{A}{S_{BET} \cdot C_c},$$

(2)

where τ is the reaction time (sec); m is the mass of catalyst (g); S_{BET} is the specific surface area data obtained from the BET-model (m²·g⁻¹); C_c is the catalyst concentration (g·L⁻¹).

Table 2: Degree of substrate conversion (C_{270} and C_{TOC}) and total (A) and specific catalytic activity (a) of investigated SO in the reaction of phenol photocatalytic degradation in water

Sample	C270, %	CTOC, %	A.10-8, M·g−1·sec−1	a.10-11, M·m−2·sec−1
-Fe2O3	—	14	0,7	0,3
WO3	—	26	1,3	2,6
ZnO	24	52	2,6	6,3
TiO2	79	85	4,2	9,2
TiO2 P25	74	91	4,5	8,6

—: the reduction of optical density does not occur.

The use of α-Fe$_2$O$_3$ and WO$_3$ samples in the reaction of phenol photocatalytic degradation leads to increase of solution optical density through the formation of colored intermediates. These oxides showed the lowest activity because of their low redox potential (Figure 1). In contrast, the usage of ZnO, TiO$_2$, and TiO$_2$ P25 samples reduces aromatic content and total organic carbon. ZnO sample takes an intermediate position among the studied semiconductor oxides by the values of the total and the specific catalytic activity. TiO$_2$ sample synthesized by a modified sol-gel method has a higher specific activity compared with TiO$_2$ P25 (9.2 and 8.6 M·m^{-2}·s^{-1}, resp.) due to predomination of crystal modification of anatase which compared with rutile has a high surface concentration of active catalytic centers [25].

Investigation of the degradation process of phenol was followed by pH measuring of the reaction mixture. In all cases the pH decrease is associated with the formation of short-chain fatty acids [26].

It is found that the specific activity of SO in the reaction of phenol photocatalytic degradation changes in a number of α-Fe$_2$O$_3$ < WO$_3$ < ZnO < TiO$_2$ P25 < TiO$_2$. Therefore, the determination of optimal conditions for phenol photocatalytic oxidation was carried out using the TiO$_2$ sample synthesized by the modified sol-gel method.

The influence of the catalyst concentration on the phenol conversion in water (Figure 5) showed a maximum efficiency at a concentration of

1 g·L⁻¹. Lower and higher concentrations of TiO₂ sample led to decrease of conversion degree associated with reduction of active sites number and radiation screening effect of TiO₂ particles excess [27].

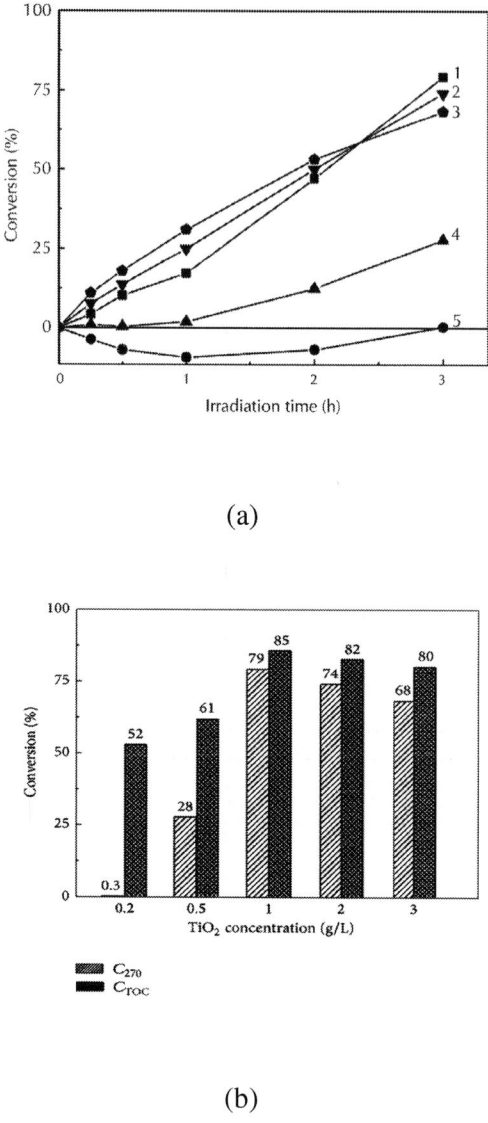

(a)

(b)

Figure 5: Effect of TiO₂ concentration on phenol conversion. Experimental conditions: initial phenol concentration = 0.536 mM; pH = 5.9;

T=303 K. (a) Variations of C_{270} values during phenol photodegradation: 1—1 g·L^{-1}; 2—2 g·L^{-1}; 3—3 g·L^{-1}, 4—0.5 g·L^{-1}, 5—0.2 g·L^{-1}; (b) C_{270} and C_{TOC} values of phenol solution after reaction.

Effect of initial phenol concentration on its conversion in water is shown on Figure 6. Rise of phenol concentration from 0.266 to 1.596 mM leads to decrease in the conversion degree for two indicators that can be attributed to an increase in the absorption of radiation by phenol molecules more than that of the catalyst particles and the increase of competitive adsorption of $^{-}$OH on the same surface site of catalyst [27].

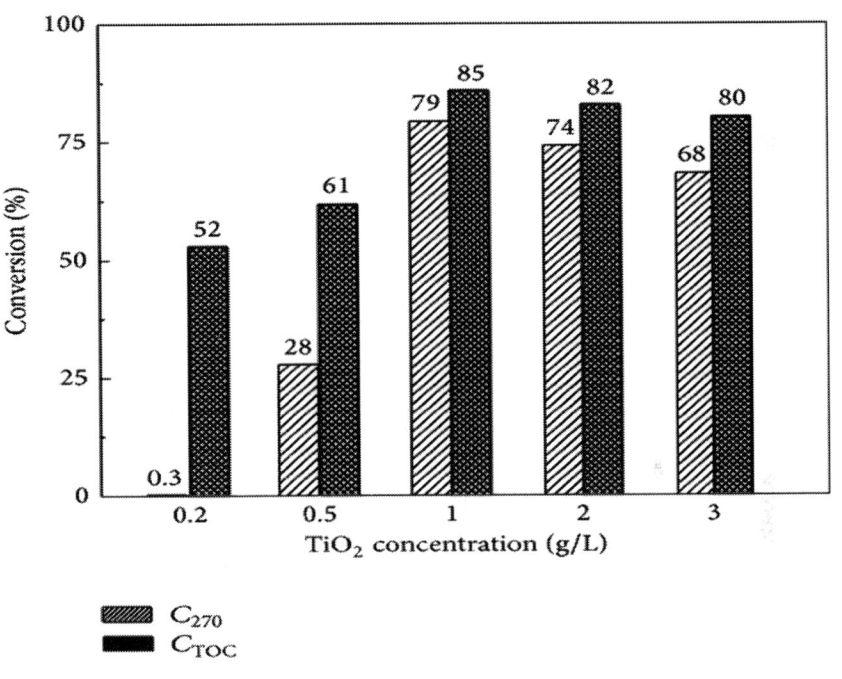

(a)

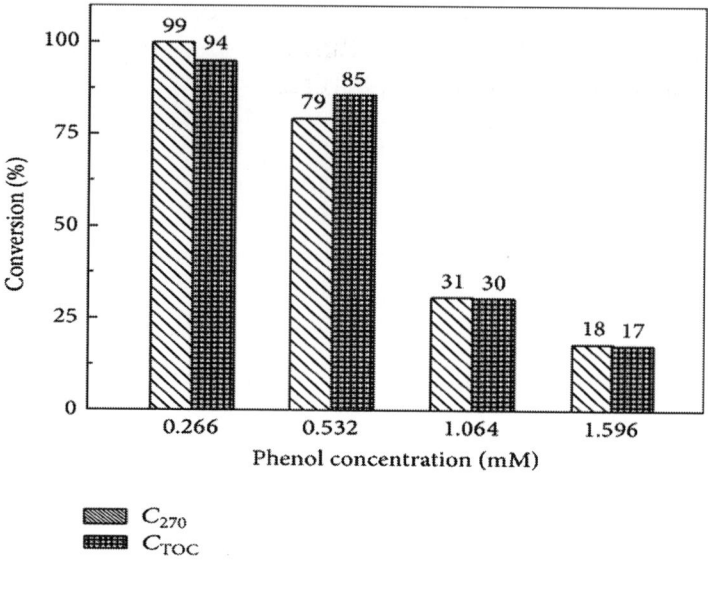

(b)

Figure 6: Effect of initial phenol concentration on its conversion. Experimental conditions: TiO_2 concentration = 0.1 g·L^{-1}; pH = 5.9; T=303K. (a) Variations of values during phenol photodegradation: 1—0.266 mM; 2—0.532 mM; 3—1.064 mM; 4—1.596 mM; (b) C_{270} and C_{TOC} values of phenol solution after reaction.

Phenol photodegradation efficiency largely depends on the pH value. Studies showed that aromatic content degradation of phenol and reduction of TOC in an acidic medium (pH = 3) are significantly low (37% and 34%, resp.) in comparison with in alkaline medium (pH = 8) (72% and 56%, resp.). Effect of pH on the photodegradation degree of phenol can be caused by changing in the surface charge of semiconductor, phenol chemical transformations in the solution, and carbonate ions formation which are effective scavengers of OH·radicals [27].

The study of photocatalytic phenol conversion dependence on reaction temperature was carried out in range from 293 to 323 K. Maximum conversion (96% of aromaticity and 86% of TOC conversion) in temperature range 303–313 K was observed.

The efficiency evaluation of the synthesized TiO_2 sample in the real conditions was carried out in Kiev tap water [28], Ukraine (Figure 7). The experimental conditions were the same as in the case of semiconductor oxides activity determination. At the beginning of the reaction rise of solution optical density takes place (Figure 7) which can be explained by the formation of intermediates. Further oxidation leads to the destruction of aromatic content and increase of conversion.

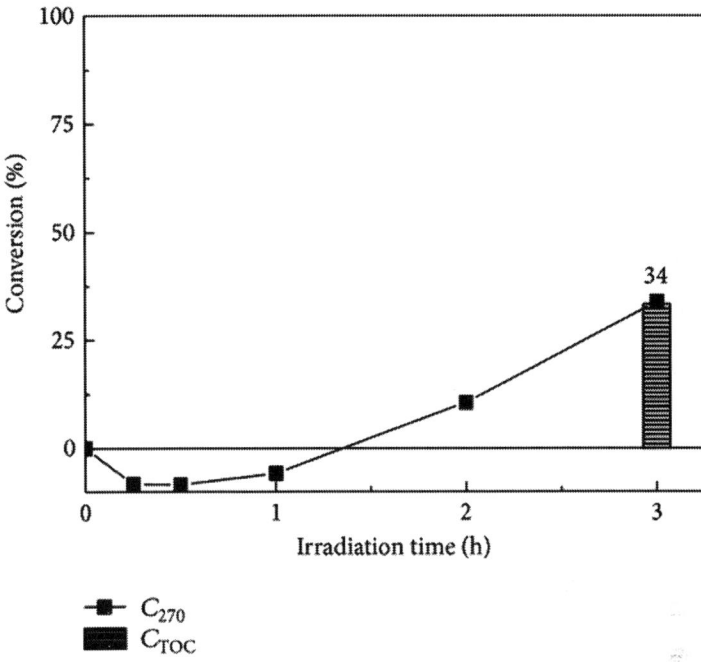

Figure 7: The photocatalytic conversion of phenol in tap water using synthesized TiO_2 samples.

CONCLUSIONS

The specific activity of investigated semiconductor oxides in the reaction of phenol photocatalytic degradation changes in a number of $\alpha\text{-}Fe_2O_3 < WO_3 < ZnO < TiO_2$ P25 $< TiO_2$. It is shown that titanium dioxide synthesized by the modified sol-gel method is the most effective photocatalyst in this process. The resulting TiO_2 has a higher specific catalytic activity than industrial sample TiO_2 P25. The maximum efficiency of the catalyst usage is achieved in 303–313 K temperature range at TiO_2 concentration of $1 g \cdot L^{-1}$, phenol concentration of 0.532 mM, and neutral pH. The presence of electrolytes reduces the efficiency of the process.

REFERENCES

1. U. I. Gaya and A. H. Abdullah, "Heterogeneous photocatalytic degradation of organic contaminants over titanium dioxide: a review of fundamentals, progress and problems," Journal of Photochemistry and Photobiology C, vol. 9, no. 1, pp. 1–12, 2008.

2. V. V. Goncharuk, Environmental Aspects of Modern Technologies Protect the Aquatic Environment, Naukova dumka, Kiev, Ukraine, 2005.

3. M. Anpo, "Utilization of TiO_2 photocatalysts in green chemistry," Pure and Applied Chemistry, vol. 72, no. 7, pp. 1265–1270, 2000.

4. D. Chen and L. Gao, "A facile route for high-throughput formation of single-crystal α-Fe_2O_3 nanodisks in aqueous solutions of Tween 80 and triblock copolymer," Chemical Physics Letters, vol. 395, no. 4-6, pp. 316–320, 2004.

5. V. V. Goncharuk, "Photocatalytic destructive oxidation of organic compounds in aqueous media,"Chemistry for Sustainable Development, vol. 5, pp. 345–355, 1997.

6. S. K. Pardeshi and A. B. Patil, "A simple route for photocatalytic degradation of phenol in aqueous zinc oxide suspension using solar energy," Solar Energy, vol. 82, no. 8, pp. 700–705, 2008.

7. N. M. Soboleva, A. A. Nosovich, and V. V. Goncharuk, "The heterogenic photocatalysis in water treatment processes," Journal of Water Chemistry and Technology, vol. 29, no. 2, pp. 72–89, 2007.

8. L. E. Sheinkman and D. V. Dergunov, "Protection of surface and groundwater from phenols pollution in underground coal mining," in Proceedings of the International Scientific and Practical Conference Science and Technology in the Modern World, 2011.

9. A. O. Samsoni-Todorov, E. A. Rolya, V. M. Kochkodan, and V. V. Goncharuk, "Photocatalytic destruction of phenol in water in the presence of cerium hydroperoxide," Journal of Water Chemistry and Technology, vol. 30, no. 3, pp. 151–156, 2008.

10. A. Memar, W. R. W. Daud, S. Hosseini, E. Eftekhari, and L. J. Minggu, "Study on photocurrent of bilayers photoanodes using different combination of WO_3 and Fe_2O_3," Solar Energy, vol. 84, no. 8, pp. 1538–1544, 2010.

11. Y. J. Jang, C. Simer, and T. Ohm, "Comparison of zinc oxide nanoparticles and its nano-crystalline particles on the photocatalytic degradation of methylene blue," Materials Research Bulletin, vol. 41, no. 1, pp. 67–77, 2006.

12. M. Grätzel, "Photoelectrochemical cells," Nature, vol. 414, no. 6861, pp. 338–344, 2001.

13. A. Martínez-de la Cruz, D. S. Martínez, and E. L. Cuéllar, "Synthesis and characterization of WO_3 nanoparticles prepared by the precipitation method: evaluation of photocatalytic activity under vis-irradiation," Solid State Sciences, vol. 12, no. 1, pp. 88–94, 2010.

14. M. Crişan, A. Brăileanu, M. Răileanu et al., "Sol-gel S-doped TiO_2 materials for environmental protection," Journal of Non-Crystalline Solids, vol. 354, no. 2-9, pp. 705–711, 2008.

15. W. Yan, H. Fan, Y. Zhai, C. Yang, P. Ren, and L. Huang, "Low temperature solution-based synthesis of porous flower-like α-Fe_2O_3 superstructures and their excellent gas-sensing properties," Sensors and Actuators B, vol. 160, no. 1, pp. 1372–1379, 2011.

16. D. Li and H. Haneda, "Morphologies of zinc oxide particles and their effects on photocatalysis,"Chemosphere, vol. 51, no. 2, pp. 129–137, 2003.

17. K. I. Gnanasekar, V. Subramanian, J. Robinson, J. C. Jiang, F. E. Posey, and B. Rambabu, "Direct conversion of TiO_2 sol to nanocrystalline anatase at 85°C," Journal of Materials Research, vol. 17, no. 6, pp. 1507–1512, 2002.

18. V. V. Goncharuk, M. V. Sychev, I. V. Stolyarova, R. V. Prihod'ko, I. O. Ledenev, and A. V. Lozovski, "MPK 7 B01 J21/00, 23/48, C 01 F1/70, The catalyst for water purification from nitrate ions,

the method of its preparation and water purification," Patent of Ukraine 7, Bulletin, 2006.

19. A. V. Tarasov, Metallurgy of Titanium, Akademkniga, Moscow, Russia, 2003.

20. A. Zecchina, G. Spoto, S. Bordiga et al., "Framework and extraframework Ti in Titanium-Silicalite: investigation by means of physical methods," Studies in Surface Science and Catalysis, vol. 69, pp. 251–258, 1991.

21. T. Sreethawong, Y. Suzuki, and S. Yoshikawa, "Synthesis, characterization, and photocatalytic activity for hydrogen evolution of nanocrystalline mesoporous titania prepared by surfactant-assisted templating sol-gel process," Journal of Solid State Chemistry, vol. 178, no. 1, pp. 329–338, 2005.

22. G. R. Bamwenda and H. Arakawa, "The visible light induced photocatalytic activity of tungsten trioxide powders," Applied Catalysis A, vol. 210, no. 1-2, pp. 181–191, 2001.

23. M. Yan, F. Chen, J. Zhang, and M. Anpo, "Preparation of controllable crystalline titania and study on the photocatalytic properties," Journal of Physical Chemistry B, vol. 109, no. 18, pp. 8673–8678, 2005.

24. Y. I. Gerasimov, "Course of Physical Chemistry," Chemistry, vol. 2, article 289, 1973.

25. M. I. Litter, "Heterogeneous photocatalysis: transition metal ions in photocatalytic systems," Applied Catalysis B, vol. 23, no. 2-3, pp. 89–114, 1999.

26. S. Ahmed, M. G. Rasul, W. N. Martens, R. Brown, and M. A. Hashib, "Heterogeneous photocatalytic degradation of phenols in wastewater: a review on current status and developments," Desalination, vol. 261, no. 1-2, pp. 3–18, 2010.

27. N. Kashif and F. Ouyang, "Parameters effect on heterogeneous photocatalysed degradation of phenol in aqueous dispersion of TiO_2," Journal of Environmental Sciences, vol. 21, no. 4, pp. 527–533, 2009.

28. V. V. Goncharuk, Kyiv Pump Room. The Quality of Artesian Water, vol. 55, Geoprint, 2003.

Resource Recovery from Industrial Effluents Containing Precious Metal Species Using Low-Cost Biomaterials — An Approach of Passive Bioremediation and Its Newer Applications

Nilisha Itankar[1], Viraja Bhat[2], Jayati Chourey[2], Ketaki Barve[2], Shilpa Kulkarni[2], Prakash Rao[2], and Yogesh Patil[3]

[1]Symbiosis Institute of Technology (SIT), Symbiosis International University (SIU), Near Lupin Research Park, Lavale, Pune, India

[2]Symbiosis Institute of International Business (SIIB), Symbiosis International University (SIU), Rajiv Gandhi Infotech Park, Hinjewadi, Pune, Maharashtra, India

[3]Symbiosis Institute of Research and Innovation (SIRI), Symbiosis International University (SIU), Lavale, Pune, Maharashtra, India

INTRODUCTION

Industrial wastes can generally be classified as wastes rich in organic matter on one hand and wastes rich in inorganic matter on the other. Cyanide (CN^-) and heavy metals (viz. copper, nickel, iron, zinc, cadmium, chromium, silver, gold, etc.) form a significant part of the latter type of wastes. Free cyanide (CN^-) is industrially important chemical because of its some unique properties of binding various transition metals to form metal-cyanides (M_xCN) complexes of variable stability and toxicity (Sharpe 1976). Therefore, cyanide finds enormous applications in variety of industrial processes. Industries like gold and silver mining, electroplating, printed circuit board manufacturing and jewellery units emanate large-volume low-tenor effluents containing anionic M_xCN complexes like gold-cyanide i.e. $[Au(CN)_2]^-$ and silver-cyanide i.e. $[Ag(CN)_2]^-$ (Vapur et al 2005). The total cyanide, gold (Au) and silver (Ag) content in these effluents ranges from 5-25, 1-2 and 5-10 mg/L, respectively (Patil 1999). The discharge limits for total cyanide is 0.2 mg/L, while for Au and Ag the standards are yet to be set and currently not available. Apart from Au and Ag many other heavy metals normally occur in the effluents in low quantity and concentrations. If inappropriately managed, cyanide and metals or their complexes can be mobilised and carried into the food web as a result of leaching from waste dumps, contaminated soils and waters. At each level of food chains, the concentration of metals increases which results into a phenomenon called biomagnification. Since cyanide is toxic and Au and Ag being precious metals, non-renewable and finite resource; their complete removal from effluents is the key.

The conventional methods adopted for the treatment of M_xCN contaminated effluents is alkaline chlorination oxidation process (Ganczarczyk et al 1985). Although this method of treatment can be very efficient in detoxifying free cyanide bearing wastes, it is not effective when challenged with anionic metal species such as M_xCN (Eckenfelder 1989). Other methods, such as copper catalyzed hydrogen peroxide oxidation, ozonation, electrolytic decomposition, etc. requires large inputs of energy, cost intensive materials and are rarely used for treatment of M_xCN containing wastes. Furthermore, at low concentration, metal-cyanide recovery by conventional means is either not possible and/or very expensive. Thus, there is a big technological breach, which needs to be bridged immediately.

Biological treatment systems (i.e. bioremediation) for the detoxification of toxic and hazardous wastes has immense potential of becoming and effective alternative because of their several advantages over conventional methods; and therefore being explored by the researchers all over the world (Patil et al., 2012; Patil and Paknikar, 1999b). However, biological methods like biodegradation / biodetoxification using live microorganisms are subject to toxicity of cyanide and metals. Therefore, removal of precious metal-cyanides species from wastes requires immediate attention of scientists and technologists. The challenge is not limited only to their removal, but also extends to finding competent and inexpensive ways of possible recovery and recycling. It was assumed that if a competent process for removal/recovery could be established, M_xCN could be conserved, which in the authors opinion would be an innovative strategy of resource recovery. Since M_xCN are anionic chemical species, therefore in principle, a well established physico-chemical methods can be used for removal and recovery of precious metal species. A few physico-chemical methods have been tried for adsorption of Au- and Ag-cyanide (Niu and Volesky 2001) and M_xCN using activated carbon or inorganic chemically active adsorbents (Lee et al 1998). However, the practical utility and cost-effectiveness of these processes are not yet established. Biosorption of metal "cations" have been studied extensively (Paknikar et al 2003). However, very few attempts have been made to adopt this technology for possible removal and recovery of "anions" such as M_xCN (Patil 2012; Patil and Paknikar 1999); especially Au- and Ag-cyanide (Niu & Volesky 2001). Literature clearly shows the paucity of references on the removal/recovery of precious Au- and Ag-cyanide using low-cost waste biomass.

It is known that biomass like bacteria, fungi, algae, plants, agricultural biomass and different agro-based industrial waste and byproducts have the ability to bind metals, in some cases selectively, from aqueous solutions (Paknikar et al., 2003). This phenomenon is named as 'metal biosorption' and the biomass responsible for the process are known as 'biosorbents'. Biosorption is combination of the processes such as electrostatic interactions, ion exchange, complexation, formation of ionic bonds, precipitation, nucleation, etc. Biomass surfaces are usually charged. The functional groups like phosphoryl, carboxyl, sulphahydryl and hydroxyl of membrane proteins, lipids and of other cell wall components are responsible for adsorption of metal (both cationic and anionic species). The overall interfaces are a result of complexity of biomass surfaces and chemical/ physical properties of metal ions (Modak and Natarajan, 1995). The advantages of passive bioremediation (i.e. biosorption) are: (i) non-living biomass is not subject to toxicity limitations; (ii) costly nutrients and aseptic conditions not required; (iii) the process is very rapid; (iv) waste biomass from fermentation and many other natural and industrial sources could be a cheap source of biomass; (v) biosorbent could be operated at wider range of pH, temperature and metal concentration; (vi) established theories, conventions and formulae could be applied to the adsorbent and many others (Paknikar et al. 2003). Although biosorption of metal cations have been studied extensively, however, very little is known about the removal and recovery of anionic metal species (viz. M_xCN) from aqueous wastes.

With the above background in mind, investigations were aimed at screening diverse range of low-cost biomass obtained from different sources for the removal and recovery of precious anionic metal species like gold- and silver-cyanide from aqueous wastes with the emphasis on the development of a laboratory / pilot scale technology.

MATERIALS AND METHODS

Low-Cost Biosorbents

The low-cost biosorbents in the present study were collected from diverse sources (as given below). Some of these biosorbents are reported for the removal of diverse metal species from acqueous solutions (Mohan & Pittman, 2006), while some of them have been employed for the first time.

- *Agricultural wastes/by-products:* Coconut fibres, Cow dung cakes, Groundnut shells, Rice husk and Rice straw
- *Industrial wastes/by-products:* Dairy waste sludge, Saw dust, Sugarcane bagasse and Tea powder waste.

- **Municipal solid waste components:** Nirmalya (waste flowers), Compost and Vegetable waste.
- **Fungal and Bacterial (waste) biomass:** *Ganoderma lucidum*, Yeast waste biomass, *Mucor heimalis*, *Penicillium* sp., *Streptomyces* waste biomass, *Streptoverticillium* waste biomass, Wood rotting fungal waste and Bacterial consortium (capable of degrading cyanide and thiocyanate). Biomass of the fungal cultures viz. *Mucor heimalis* and *Penicillium* sp. available in our laboratory were obtained by cultivating them aseptically in Sabouraud's medium (Composition: Glucose 20 g; Peptone 10 g; Distilled water 1000 ml; pH 4.5-5.0) for 4-5 days at 30°C on rotary shaker incubator (150 rpm). Biomass was harvested after growth by filtering through muslin cloth and washed with distilled water 3-4 times in order to remove the organic traces. After washing, the biomass was subjected to drying in oven at 50-60°C for 2-3 days till the constant weight was obtained. Biomass was then ground using electric mixer in order to obtain particle size of ≤ 500 µm (0.5 mm). Later the biomass were stored in glass bottles with suitable air tight caps for further use.
- **Algae:** Mixed algae biomass obtained from lake Rankala, Kolhapur
- **Terrestrial and aquatic plant species:** *Parthenium* sp., *Eichhornia* root biomass, *Eichhornia* stem biomass, *Eichhornia* leaf biomass, Runners, *Tectona grandis* waste leaves and *Lantana camara*.
- **Reference material:** Activated charcoal was employed as a reference material in order to obtain comparative data.

Biomass samples were collected in polythene bags and transported to laboratory. The samples were washed several times with tap water to remove the dirt and other contaminants, if any, and was then finally washed with deionised water (< 5 µS). Biomass were then subjected for drying at 50°C for 48-72 h to a constant weight and powdered. Dried biomass was pulverized employing electric mixer and sieved; so as to get uniform particle size of ≤ 500 µm (0.5 mm).

Synthesis of Stock Anionic M$_x$CN Solutions

The stock solutions of Au-cyanide i.e. $[Au(CN)_2]^-$ (Dicyanoaurate-DCAU) and Ag-cyanide i.e. $[Ag(CN)_2]^-$ (Dicyanoargentate - DCAG) were prepared stoichiometrically by combining their respective salts with sodium cyanide in the molar proportion of 1:2 (Patil & Paknikar, 2000a; Patil and Paknikar,2000b; Rollinson et al., 1987). Spectral properties were checked and confirmed periodically using UV spectrophotometer. The synthesized DCAU and DCAG solutions were refrigerated at 8-10°C.

Chemicals and Analyses

Chemicals used for all experiments were of analytical grade (AR). Glassware used were made of borosilicate material. Stock solutions and reagents were prepared in deionized water (< 5 µS) and stored in refrigerator (8-10°C). Gold (Au), silver (Ag), nickel (Ni), copper (Cu), zinc (Zn) and iron (Fe) in the experimental solutions and effluents were analysed using Atomic Absorption Spectrophotometer (Elico, India SL-173). Total cyanide and chemical oxygen demand (COD) content in the solutions were estimated by pyridine-barbituric acid and reflux method, respectively as described in Standard Methods (APHA-AWWA-WEF, 1998). Phosphates (PO_4^{-3}) from effluents were analysed by phenol- disulphonic acid method; sulphates (SO_4^{-2}) were determined by barium chloride method while chlorides (Cl^-) were determined by argentometric method, as per the methods prescribed in Standard Methods (APHA-AWWA-WEF, 1998). Colour and turbidity were recorded by visual observations. pH and electrical conductivity from solutions was measured by their respective meters.

In order to determine the inherent/actual pH of each powdered unconditioned biomass, the biomass sample and RO water were mixed and serially diluted in the ratio of 1:20, 1:30, 1:40 and 1:50 (w/v) in conical flasks. The contents were stirred vigourously and kept for one hour in stationary conditions and was followed by determining the pH of each dilution ratio. pH value obtained for each dilution was then plotted on graph of "pH against water-to-biomass ratio (v/w)". The straight line obtained after joining all the points was extrapolated backwards so as to intersect with Y-axis (i.e. the pH scale).

Gold-Cyanide (DCAU) and Silver-Cyanide (DCAG) Biosorption Studies

A batch equilibration method was used to determine the sorption of DCAU (0.02 mM i.e. 3.94 mg/l Au and 2.08 mg/l CN⁻) and DCAG (0.1 mM i.e. 10.78 mg/l Ag and 5.2 mg/l CN⁻). Biosorbent (0.05 to 0.2 g) was contacted with 10 ml solution of DCAU or DCAG of desired pH in a set of 50 ml capacity conical flasks. The flasks were incubated on rotary shaker incubator adjusted to a speed of 150 rpm at 30°C for 1 h. Contents of flasks were filtered using ordinary filter paper and then analysed for residual Au, Ag and cyanide. All experiments were performed in duplicates and repeated twice to confirm the results. Appropriate controls were run simultaneously to detect the air stripping of cyanide, if any, to confirm biosorption.

Influence of pH on biosorption of DCAU/DCAG was checked in the range of 4-10 with pre-conditioned biomass. On the basis of maximum DCAU/DCAG uptake values obtained under optimum pH conditions, efficient biosorbents were selected for further studies. DCAU/DCAG loading capacity (µmol DCAU/DCAG bound per gram weight of biosorbent) of each biosorbent was determined by contacting 0.1 g powdered biomass several times with fresh batches of 10 ml DCAU/DCAG solution till saturation was achieved. To determine optimum biosorbent amount, DCAU/DACG was contacted with varying amounts of biomass powder, ranging from 0.1 to 5% (w/v). Rate of DCAU/DCAG uptake was studied by contacting the biosorbent for a period ranging between 0 to 5 h. Under optimised conditions, effect of various competing cations viz. Cu^{2+}, Ni^{2+}, Zn^{2+}, Cd^{2+}, Pb^{2+}, Fe^{2+}, Ag^+, etc. (0.01-0.1 mM) and anions viz. SO_4^{2-}, NO_3^-, Cl^-, PO_4^{3-}, etc. (0.1-1 mM) on biosorption of DCAU/DCAG was also checked. In order to test the effect of pre-treated biomass on uptake of DCAU/DCAG, the biosorbent were treated for one hour using L-cysteine, boiling water, sodium hydroxide, formaldehyde, acetone, acetate, methanol and acetic anhydride prior to sorption.

Adsorption Isotherms

To study the impact of initial concentration on adsorption, varying concentration of DCAU/DCAG was used in the range of 0.01 to 1 mM.

In order to obtain sorption data, uptake value (Q) was calculated using the following equation:

$$Q = V \left(C_i - C_f \right) / 1000 \text{ m}$$

(1)

Where, Q is DCAU/DCAG uptake (mmol per gram biomass); V is the volume of DCAU/DCAG solution (ml); C_i is the initial concentration (μmol); C_f is the final concentration (μmol); m is mass of sorbent (g). Based on the 'Q' value obtained adsorption isotherms were plotted according to Freundlich and Langmuir equations (Freundlich, 1926; Langmuir, 1918):

$$\ln Q = \ln K + \left(1/n \right) C_{eq} \text{ Freundlich equation}$$

2)

$$C_{eq} / Q = \left(1 / b \, Q_{max} \right) + \left(C_{eq} / Q_{max} \right) \text{ Langmuir equation}$$

(3)

Where, C_{eq} is the liquid phase concentration of DCAU/DCAG; b is Langmuir constant; Q_{max} is maximum DCAU/DCAG uptake; K is constant; n is the number of metal reactive sites and Q is the specific metal uptake.

Adsorption/Desorption of DCAU and DCAG

Samples of 1 g biosorbent loaded with target $M_x CN$ was eluted using desorbing agent (1-3 N NaOH) in concentrated form and analysed. Following the elution, biosorbent was washed with DW and then again conditioned to appropriate optimum pH to use in next adsorption/desorption cycle.

Biosorption of AU- and AG-Cyanide from Industrial Wastewaters

Two types of effluents were procured from silver and gold plating industry. Both effluents were subjected to characterization using Standard methods (APHA-AWWA-WEF, 1998). The proximate analysis of the samples is shown in Table 8 and 9. Batch equilibration method was followed as mentioned earlier. Rice husk (0.1 g) and *Eichhornia* root (0.1 g) biomass was contacted with 10 ml of gold-cyanide and silver-cyanide effluents, respectively. Prior to sorption, the gold- and silver-cyanide effluents were adjusted to desired optimum biosorption pH. All the batch sorption experiments were carried out under optimum conditions as given in Table 1. After contact, the contents of the flasks were filtered and then analysed for residual metal (i.e. gold and silver) and cyanide. Appropriate controls were run simultaneously.

Table 1: Optimum conditions used for biosorption experiments

Parameters	For DCAU experiments	For DCAG experiments
Biomass	**Rice husk (L-cysteine treated)**	***Eicchornia* root biomass (L-cysteine treated)**
pH	4.0	6.0
Temperature (°C)	30	30
Biomass quantity (w/v)	1.0	2.0
Contact time (min)	60	60
Rotation speed (rpm)	150	150

Continuous Biosorption Studies Using Fixed Bed Column at Laboratory Level

Scale-up studies in fixed bed continuous mode at laboratory level for biosorption of gold- and silver-cyanide was carried out in two separate fabricated glass columns of height 44 cm, internal diameter 1.3 cm and filter media height being 30 cm. The total volume of the column was 58.37 cm³, while the working volume was 39.80 cm³ (figure not shown). Glass column no. 1 was filled with 21 g rick husk biomass pretreated with L-cysteine, while the glass column no. 2 was filled with 24 g *Eicchornia* root biomass also pretreated with L-cysteine. The target effluents were passed through the columns in upward direction in continuous mode at a flow rate of 40 ml/h using programmable peristaltic pump (Enertech-Victor, India). All connecting silicon tubings used in the experiments were of 0.5 cm outer diameter and 0.3 cm inner diameter. Gold-cyanide effluent was passed through the column no. 1 upto 50 bed volumes, while silver-cyanide effluent was passed upto 34 bed volumes till the breakthrough curve (S-shaped) was obtained. Samples were collected periodically after every two hours and analysed for Au, Ag and total cyanide content.

Biodegradation of Residual M_xCNS

Unrecoverable (residual) gold-cyanide and silver-cyanide in the solutions after biosorption treatment (in batch studies) were subjected to biodegradation using "live chemoheterotrophic bacterial consortium". The consortium (comprising of three *Pseudomonas* sp. in a proportion of 1:1:1) capable of degrading free cyanide and thiocyanate as the source of nitrogen was isolated by enrichment culture technique (Patil, 2008) and was available in my laboratory. Biodegradation experiment were conducted under aerobic and optimum conditions of pH (7.0), temperature (30°C), inoculum size (10^7 cells/ml) and glucose concentration (1 mM). The biodegradation process was used as a polishing step to clean the effluent containing traces of cyanide in order to meet the requirements of statutory agencies.

RESULTS

Screening of Low-Cost Waste Biomass for DCAU and DCAG Sorption

Data in Table 2 summarizes the results obtained for DCAU and DCAG sorption under optimal pH conditions. The results showed that optimum sorption in terms of Q (i.e. μmol M_xCN sorbed per gram biomass) of 0.02 mM DCAU and 0.1 mM DCAG for most of the waste biomass/sorbents tested were at pH 4.0 and 6.0, respectively. It was observed that biosorption of DCAU and DCAG was less above pH 7.0 for all the biomass tested. In acidic pH conditions, sorption of DCAU and DCAG increased significantly. The table also shows that other than activated charcoal (chosen as reference material) which showed highest biosorption capacity, biomass of Rice husk (3.65 µmol/g) and *Eichhornia* roots (3.56 µmol/g) were efficient biosorbents for DCAU sorption; while *Eichhornia* roots (4.76 µmol/g) and Tea powder waste (4.73 µmol/g) were efficient biosorbents for DCAG. The overall Q values observed for all the waste sorbents tested for DCAU and DCAG were in the range of 2.69 - 3.65 µmol/g and 2.74 - 4.76 µmol/g, respectively. The observed Q values for efficient biomass were found to be marginally below the Q values obtained for activated charcoal (3.80 - 5.00 µmol/g). As far the optimum pH for sorption was concerned, DCAU uptake was maximal at pH 4.0 for all the biomass tested, while DCAG uptake for majority of the biomass was at pH 5.0 to 6.0. There was no loss of DCAU or DCAG in the control flasks without sorbent during the tested time period.

Table 2 also shows the data on pH values of all unconditioned biomass. Other than the reference materials, the lowest pH observed was that of coconut fibers (pH 4.24), while the highest pH was of mixed algae biomass (pH 7.61). pH of unconditioned Rice husk, Tea powder waste and *Eichhornia* root biomass observed were 5.94, 4.94 and 7.01, respectively, while their optimum pH of biosorption was 4.0 (for DCAU biosorption), 5.0 - 9.0 (for DCAG biosorption) and 4.0 (for DCAU biosorption) and 7.0 – 9.0 (for DCAG biosorption).

On the basis of maximum DCAU/DCAG uptake values obtained under optimum pH conditions, Rice husk and *Eichhornia* root biomass were selected for DCAU sorption, while *Eichhornia* root and Tea powder waste biomass were selected for DCAG sorption for further experiments. Activated charcoal acted as a reference material.

Influence of Temperature on Biosorption of DCAU and DCAG

It was observed that biosorption of DCAU and DCAG by the selected biomass did not had any significant impact with the change in temperature of the system from 5-45°C.

DCAU and DCAG Loading Capacity

Table 3 and 4 depicts the data on DCAU and DCAG loading capacity of pre-conditioned (at pH 4.0 for DCAU and pH 5.0-7.0 for DCAG) biosorbents selected on the basis of maximum sorption under optimum pH, as described earlier. Also the results were compared with the unconditioned biomass (i.e. the original pH of the biomass itself). It could be seen that Rice husk biomass had the maximum loading capacity for DCAU (7.63 µmol/g) sorption among the two tested biomass; and was followed by Eichhornia root biomass (7.04 µmol/g). It was also observed that the loading capacity of activated charcoal was found relatively lower when compared with the Rice husk. While the loading capacity of unconditioned biomass dropped by 7.6% and 43% for Rice husk and Eichhornia root biomass, respectively.

In case of DCAG, Eichhornia root biomass showed highest loading capacity (9.74 µmol/g) followed by Tea powder waste (9.41 µmol/g). Loading capacity values of Eichhornia root biomass was highly competitive and comparable with activated charcoal (9.95 µmol/g), which was used as reference material. Furthermore, the loading capacity of unconditioned biomass was not affected when compared with the conditioned biomass (Table 4)

Table 2: Biosorption of DCAG and DCAU at optimum pH

Sr. No.	Biosorbent	pH of unconditioned biomass	Q (µmol M_xCN sorbed per gram biomass)	
			DCAU	DCAG
(A)	**Agricultural waste/by-products**			
1.	Coconut fibers	4.24	3.32 (4.0)*	4.62 (5.0)*
2.	Cow dung cakes	7.73	3.07 (4.0)	4.64 (6.0)
3.	Groundnut shells	5.49	3.22 (4.0)	4.62 (6.0)
4.	Rice husk	5.94	3.65 (4.0)	4.68 (6.0)
5.	Rice straw	6.13	3.25 (4.0)	4.69 (6.0)
(B)	**Industrial waste/by-products**			
6.	Dairy waste sludge	6.88	2.91 (4.0)	4.71 (6.0)
7.	Saw dust	5.59	3.52 (4.0)	4.64 (6.0)
8.	Sugarcane Bagasse	5.92	3.16 (4.0)	4.03 (6.0)
9.	Tea powder waste	4.94	2.94 (4.0)	4.73 (5.0-9.0)

(C)	Municipal solid waste components			
10.	Nirmalya (Waste flowers)	6.20	3.43 (4.0)	4.60 (6.0)
11.	Compost	7.28	3.09 (4.0)	3.22 (6.0)
12.	Vegetable waste	6.77	2.91 (4.0)	3.88 (6.0)
(D)	Fungal and Bacterial waste/biomass			
13.	Ganoderma sp.	6.04	3.01 (4.0)	4.07 (6.0)
14.	Yeast biomass	4.39	2.69 (4.0)	3.06 (5.0)
15.	Mucor heimalis	4.45	1.97 (4.0)	2.29 (6.0)
16.	Penicillium waste	4.26	3.08 (4.0)	3.99 (6.0)
17.	Streptomyces waste	4.86	3.00 (4.0)	2.78 (6.0)
18.	Streptoverticilliumwaste	4.67	2.77 (4.0)	3.52 (6.0)
19.	Wood rotting fungi	6.04	3.18 (4.0)	4.21 (6.0)
20.	Bacterial consortium	6.83	3.12 (4.0)	4.00 (6.0)
(E)	Algae biomass			
21.	Mixed algae biomass	7.61	3.29 (4.0)	4.16 (6.0)
(F)	Photosynthetic trees/plants waste			
22.	Parthenium sp.	6.69	3.07 (4.0)	4.22 (6.0)
23.	Eichhornia leaves	5.57	3.20 (4.0)	4.62 (6.0)
24.	Eichhornia roots	7.01	3.56 (4.0)	4.76 (7.0-9.0)
25.	Eichhornia stem	5.58	3.38 (4.0)	4.66 (6.0)
26.	Runners	6.52	3.06 (4.0)	4.67 (6.0)
27.	Tectona grandis leaves	5.40	3.42 (4.0)	4.63 (6.0)
28.	Lantana camara leaves	6.59	2.98 (4.0)	2.74 (5.0)
(G)	Reference materials			
29.	Activated charcoal	5.59	3.80 (4.0)	5.00 (6.0)
30.	Bagasse Fly ash	8.75	3.40 (4.0)	4.70 (6.0)
	Control (without biomass)	-	0 (2.0)	0 (6.0)

[i] - All the values in table are average of two readings; *Values in parentheses indicates optimum pH (Gaddi and Patil, 2011; Patil, 2012)

Table 3: DCAU loading capacity of selected biosorbents

Sorbent / Biosorbent	Loading capacity (µmol/g of biomass)		
	Conditioned biomass (at optimal pH)	Unconditioned biomass (at original biomass pH)	Remarks
Rice husk	7.63 (4.0)	7.05 (5.94)	Moderately affected

Eichhornia roots	7.04 (4.0)	4.38 (7.01)	Significantly affected
Activated charcoal	7.61 (4.0)	7.58 (5.59)	Not affected

[i] - All the values presented in table are average of two readings

Table 4: DCAG loading capacity of selected biosorbents

Sorbent / Biosorbent	Loading capacity (µmol/g of biomass)		
	Conditioned biomass (at optimal pH)	Unconditioned biomass (at original biomass pH)	Remarks
Eichhornia roots	9.74 (7.0)	9.77 (7.01)	Not affected
Tea powder waste	9.41 (5.0)	9.40 (4.94)	Not affected
Activated charcoal	9.95 (6.0)	9.94 (5.59)	Not affected

[i] - All the values presented in table are average of two readings

Considering the above results, selection of the biosorbent was further narrowed down to Rice husk and*Eichhornia* root biomass for DCAU and DCAG biosorption, respectively, (using conditioned biomass) for further experiments.

Influence of Biosorbent Quantity

The effect of biomass quantity (% w/v) on DCAU and DCAG biosorption was studied at optimal pH values. Varying amount of biomass ranging from 0.1 to 5.0 g were used keeping the volume of both the metal-cyanides (M_xCNs) solution constant (10 ml); thereby giving the solid-to-liquid ratio in the range of 0.01 to 0.5. The results showed that the biomass quantity increased the % biosorption of both DCAU and DCAG also increased. Maximum uptake in terms of Q (3.84 µmol/g) was observed at 1% (w/v) of Rice husk biomass for DCAU sorption. However, from 1 to 5 % (w/v) there was no significant increase. In case of DCAG sorption, *Eichhornia* root biomass showed highest Q for biomass quantity from 2.0 to 5.0% (w/v).

Rate of DCAU and DCAG Uptake

The effect of contact time on DCAU and DCAG biosorption was studied at their optimum pH (pH 4.0 and pH 6.0 for DCAU and DCAG sorption, respectively), temperature (30°C) and biomass quantity of 1% (w/v) and 2% (w/v) for DCAU and DCAG, respectively. 10 ml of precious M_xCN solution having concentration 0.02 mM (in case of DCAU) and 0.1 mM (in case of DCAG) was contacted with respective biomass (Rice husk and *Eichhornia* root biomass for DCAU and DCAG, respectively) for the period upto 180 min. The time intervals chosen for study were 0 to 180 minutes. Periodically the flask contents were removed by filtration and the filtrates were analyzed for Au, Ag and cyanide concentration.

It was observed that rate of both the M_xCN uptake was maximum in the 15-20 minutes, with over 80% of biosorption. Later, the sorption rate slowed down until it reached a plateau after 30-40 min, indicating equilibration of the system. Maximum sorption of both the precious M_xCNs was observed at 40 min (88.2% for DCAU and 94.3% in case of DCAG).

Adsorption Isotherm Models

The effect of initial concentration provides an important driving force to overcome all mass transfer resistance of target inorganic ion between the aqueous and solid phases. The biosorption of both DCAU and DCAG were carried out at different initial concentrations ranging from 0.01 mM to 1.0 mM (corresponding to approximately 10 to 1000 μmol) at pH 4.0 and 6.0 using 1 and 2% (w/v) of Rice husk and *Eichhornia* root biomass, respectively. It was found that the equilibrium sorption capacity of the sorbent increased with increasing initial concentration of M_xCNs from 0.01 mM to 1.0 mM, due to the increase in the number of ions competing for the available binding sites in the biomass. The uptake of M_xCNs approached towards plateau above 0.5 mM. There was a significant increase in the specific uptake of both M_xCNs

Equilibrium Models to Fit Experimental Data

To examine the relationship between sorption isotherm models are widely employed for fitting the data. Langmuir and Freundlich were used to describe the equilibrium between the two M_xCNs sorbed on Rice husk and *Eichhornia* root biomass and M_xCNs in solution. Data obtained show that M_xCNs uptake values could be well fitted to the Langmuir and Freundlich isotherm models with the regression value >0.98.

Influence of Cationic and Anionic Moieties on DCAU and DCAG SORPTION

It was observed that biosorption of both the metal-cyanides were not significantly affected by the presence of various metal cations and anions in majority of the cases. Biosorption of DCAU was affected by zinc, chromium and cadmium up to certain extent (33-40%). In case of DCAG, biosorption was affected significantly by the presence of cadmium, zinc, iron and chromium (37-67%). Biosorption in the presence of other metals cations (copper, nickel and silver) and anions (phosphates, sulphates and chlorides) was consistently above 80%.

When the low-cost biomass was pre-treated with different chemicals, it was found that (Table 5 and 6) there was greater degree of variation in the biosorption of DCAU and DCAG using Rice husk and*Eichhornia* root biomass, respectively. Rice husk and *Eichhornia* root biomass treated with 1% L-cysteine enhanced the biosorption capacity of both the M_xCNs. In contrast, the NaOH pretreated biomass significantly hampered the biosorption process. It was 0% in case of DCAU sorption and 2.5% in case of DCAG biosorption.

Table 5: Impact of pretreatment on DCAU biosorption by Rice husk

DCAU + Chemicals used for pre-treatment	% DCAU biosorption	Relative % biosorption
DCAU without sorbent (control)	0	0
DCAU + Rice husk without pretreatment	90.1	100
DCAU + Rice husk (treated with boiled water)	86.5	96.0

DCAU + Rice husk (treated with 1% L-cysteine)	100	110.9
DCAU + Rice husk (treated with 1 N NaOH)	0	0
DCAU + Rice husk (treated with 1 N Formaldehyde)	87.2	96.7
DCAU + Rice husk (treated with acid i.e. HCl)	79.0	87.7
DCAU + Rice husk (treated with acetate)	73.6	81.7
DCAU + Rice husk (treated with methanol)	27.1	30.1
DCAU + Rice husk (treated with acetic anhydride)	49.2	54.6
DCAU + Rice husk (treated with acetone)	53.7	59.6

The experiment on pretreatment of biomass with L-cysteine clearly showed enhanced biosorption of DCAU and DCAG from solutions. It was therefore thought worthwhile to find out the loading capacity of both the biomass pretreated with L-cysteine. Experiment on loading capacity of Rice husk and Eichhornia root biomass was performed as mentioned earlier (section 2.4). It could be seen from Table 7 that the loading capacity of Rice husk and Eichhornia root biomass enhanced the biosorption of DCAU and DCAG up to 175% and 140%, respectively compared to untreated biomass (i.e. in absence of L-cysteine loaded biomass).

TABLE 6: Impact of pretreatment on DCAG biosorption by Eichhornia root biomass

DCAU + Chemicals used for pre-treatment	% DCAG biosorption	Relative % biosorption
DCAU without sorbent (control)	0	0
DCAU + Eichhornia root without pretreatment	94.0	100
DCAU + Eichhornia root (treated with boiled water)	83.0	88.3

DCAU + Eichhornia root (treated with 1% L-cysteine)	100	106.4
DCAU + Eichhornia root (treated with 1 N NaOH)	2.5	0.5
DCAU + Eichhornia root (treated with 1 N Formaldehyde)	89.4	95.1
DCAU + Eichhornia root (treated with acid i.e. HCl)	81.7	86.9
DCAU + Eichhornia root (treated with acetate)	64.9	69.0
DCAU + Eichhornia root (treated with methanol)	34.5	36.7
DCAU + Eichhornia root (treated with acetic anhydride)	56.6	60.2
DCAU + Eichhornia root (treated with acetone)	59.0	62.7

Table 7: Loading capacity of untreated and L-cysteine treated biomass

Biosorbent	Loading capacity (μmol/g of biomass)	
	Biomass without pretreatment	L-cysteine treated biomass
Rice husk (for DCAU biosorption)	7.60 (100%)	13.34 (175%)
Eichhornia roots (for DCAG biosorption)	9.72 (100%)	13.62 (140%)

[i] - All the values presented in table are average of two readings

Adsorption-Desorption of DCAU and DCAG

The loaded DCAU and DCAG on Rice husk and *Eichhornia* root biomass, respectively, could be desorbed with more than 97% efficiency using 1 N sodium hydroxide solution. Final concentrations of M_xCNs in the eluent were 28-30 folds of initial concentration of DCAU and 22-25 fold of the initial concentration of DCAG. However, during the second cycle of M_xCN adsorption, the loading capacity of the biosorbent decreased by 10-15%.

Biosorption of DCAU and DCAG from Industrial Wastewaters

The gold-cyanide and silver-cyanide from the effluents procured from the industries could be effectively biosorbed/treated by Rice husk and *Eichhornia* root biomass which were pretreated with L-cysteine. Table 8 and 9 depicts the data on gold-cyanide and silver-cyanide before and after biosorption along with their percentage removal. Gold and cyanide removal efficiency from gold-cyanide effluent was 91.53% and 82.69%, respectively. However, the cyanide content in the treated effluent after biosorption although very less (0.59 and 0.78 mg/l for Au- and Ag-cyanide effluents, respectively) but was not complying with the prescribed Indian Sstandards (0.2 mg/l). Overall, the results indicated that both Rice husk and *Eichhornia* root biomass were very effective in treating the effluents by biosorption process.

TABLE 8: Biosorption of gold-cyanide from industrial effluent in batch mode using rice husk pretreated with L-cysteine

Physicochemical parameters	Before biosorption	After biosorption	% Removal efficiency
Color	Colorless	Colorless	-
Turbidity	Clear	Clear	-
pH	6.87	4.12	-
Total cyanide	3.41	0.59	82.69
Gold	1.30	0.11	91.53
Silver	0.48	0.03	93.75
Copper	0.95	0.18	81.05

Nickel	BDL	-	-
Zinc	0.50	0.10	80.00
Iron	0.11	BDL	100.00
Phosphates	97.9	76.1	22.26
Sulfates	63.5	61.3	3.46
Chlorides	173.0	155.2	10.28
Chemical oxygen demand (COD)	42	31	26.19

[i] - All the figures given in the table are in mg/l, except pH; BDL: Below detectable limits

Table 9: Biosorption of silver-cyanide from industrial effluent in batch mode using *Eichhornia* root biomass pretreated with L-cysteine

Physicochemical parameters	Before biosorption	After biosorption	% Removal efficiency
Color	Colorless	Colorless	-
Turbidity	slightly turbid	Clear	-
pH	7.16	6.44	-
Total cyanide	5.02	0.78	85.0
Gold	-	-	-
Silver	7.29	0.98	86.55
Copper	1.56	0.31	80.12
Nickel	BDL	-	-
Zinc	0.92	0.27	70.65
Iron	0.18	BDL	100.0
Phosphates	117.5	115.0	2.12
Sulfates	94.1	95.2	0
Chlorides	199.6	193.3	3.15
Chemical oxygen demand (COD)	47	29	38.29

[i] - All the figures given in the table are in mg/l, except pH; BDL: Below detectable limits

In order to treat the residual (unrecoverable) cyanide remaining in the solutions after biosorption were subjected to biodegradation process using bacterial consortium under optimized conditions in further experiments.

Biodegradation of Unrecoverable (Residual) Metal-Cyanides

Typical residual concentrations of gold and cyanide in gold-cyanide effluent after biosorption were 0.11 and 0.59 mg/l, respectively. Similarly, the residual silver and cyanide concentration in the silver-cyanide effluent were 0.98 and 0.78 mg/l, respectively. When these solutions were subjected to biodegradation using bacterial consortium under optimum conditions as mentioned earlier, it was observed that the consortium could degrade the said cyanide from both effluents with an efficiency exceeding 90% within a period of 3-4 h. pH, cyanide and chemical and oxygen demand (COD) of the treated effluent were within the permissible limits prescribed by statutory agencies in India (Table 10). Percent cyanide removal efficiency was >90% for both types of effluents. Gold and silver metals were not detected in bacterial free treated solutions. Findings indicated that biodegradation could be used as a polishing step in the treatment of M_xCNs containing wastewaters.

TABLE 10: Treatment of residual gold-cyanide and silver-cyanide by a cyanide and thiocyanate degrading heterotrophic bacterial consortium

Parameter	Biodegradation		*BIS Standards	% Removal Efficiency
	Before	**After**		
Gold-cyanide				
pH	6.99-7.04	7.05-7.12	5.5 – 9.0	-
Total cyanide	0.59 mg/l	0.04	0.2	93.22%
Chemical oxygen demand (COD)	102 mg/l	23 mg/l	250	77.45%
Gold	0.11	BDL	NA	-
Silver-cyanide				
pH	6.95-7.03	7.07-7-11	5.5 – 9.0	-

Total cyanide	0.78 mg/l	0.05	0.2	93.58%
Chemical oxygen demand (COD)	98 mg/l	19 mg/l	250	80.61%
Silver	0.98	BDL	NA	-

[i] - *BIS- Bureau of Indian Standards; BDL-Below Detectable Limits; NA-Not Available; All the values presented in table are average of two readings

Biosorption of Gold- and Silver-Cyanide Effluent in Packed Bed COLUMN

Biosorption studies on gold-cyanide and silver-cyanide effluents were performed in continuous mode in two separate packed bed glass columns consisting of Rice husk (column 1) and *Eichhornia* root biomass (column 2), respectively. Biosorption results showed that breakthrough point observed for gold and cyanide in column 1 was 60 h, while the breakthrough time observed for silver and cyanide in column 2 was 40 h (figures not shown). The total effluent passed through the column 1 and 2 was equivalent to 50 and 34 bed volumes, respectively. Column 1 and 2 got completely saturated after 90 and 70 h, respectively.

DISCUSSION

Review of literature show that biosorption of heavy metal cations from aqueous solutions have been studied widely (Paknikar et al., 2003). Studies have also been carried out on biosorption of anionic metal species like chromates (Basha et al., 2008; Itankar and Patil, 2012), free cyanide (Azab et al., 1995) and metal-cyanides (Patil and Paknikar, 1999a) using microbial biomass, especially the waste fungal biomass obtained from fermentation industry and laboratory cultivated biomass. In contrast, biodetoxification of metal-cyanides and thiocyanate using live bacterial consortium was also studied (Patil, 2006; Patil, 2008a; Patil, 2008b; Patil, 2011; Patil and Paknikar, 2000; Patil and Paknikar, 2001). Safety aspects of cyanide use in mining industries have been well emphasized by Patil and Kulkarni (2008). Prashanth and Patil (2007) have also studied the impact of free cyanide on edible fish *Catla catla*.

Another important and precious chemical species that are normally encountered in the industrial effluents emanating from mining, electroplating, printing circuit board manufacturing, photography units, etc. are gold-cyanide and silver-cyanide. These species are active and important members of cyano-group chemicals that occur in water environment. Some research has been carried out on the removal of gold-cyanide and silver-cyanide species by biodegradation/biodetoxification method (Karavaiko et al. 2000; Kiruthika and Shrinithya, 2008). However, very little information is available on the removal and recovery of gold-cyanide and silver-cyanide from high volume low tenor effluents (Gaddi and Patil, 2011). Much work has been restricted to the removal of metal-cyanides and thiocyanates using anion exchange resins and activated charcoal (Kononova et al., 2007) and polyurethane foam (Hasany, 2001). Some papers on removal of free thiocyanate and metal-thiocyanate have also been published using low-cost materials (Namasivayam, 2007; Thakur and Patil, 2009). Overall literature survey shows that very little work has been carried out on the removal of anionic species like Au-cyanide i.e. $[Au(CN)_2]^-$ (Dicyanoaurte-DCAU) (Niu and Volesky, 2000) and Ag-cyanide i.e. $[Ag(CN)_2]^-$ (Dicyanoargentate-DCAG) (Gaddi and Patil, 2011) from waste solutions using low-cost materials emanated either by natural or manmade activities. Since all the cyano-group chemicals like free cyanide and thiocyanate, metal cyanides and metal thiocyanate are toxic to all classes of living cells their removal and recovery from waste prior to discharge in environment is the key.

In the light of above, the present research work was focused to study removal and recovery of Au- and Ag-cyanide from effluents using low-cost biosorbents (using waste biomass from various sources); followed by biodegradation (using active bacterial consortium). It was contemplated that if an efficient process for removal and recovery could be developed, then precious M_xCN or metals could be conserved, which according to the project investigators opinion would be an innovative approach of resource recovery.

It is well-known that certain type of microbial or waste biomass has high degree of competency to adsorb heavy metals. This sorption is solely due to the chemical composition of biomass (Volesky, 2003). With biosorption applications in mind it makes sense to screen variety of biomass types that are readily available in large quantities. There are basically two types of biomass sources that can practically be considered with low costs and availability in mind. First, the industrial waste biomass generated as a by-product of large scale (for example fermentation industry) with virtually no uses for it and disposal is a problem. Secondly, the biomass generated in large quantities from water environment (for examples unwanted plants like *Eichhornia* sp. and algae). It can be easily collected or harvested as raw material for biosorbents. Also, there are many other sources from where low-cost biomass could be procured especially in a developing country like India. These include the vegetable waste, yard wastes, waste flowers and coconut fibres from temples, etc. Energy generation potential from biomass and MSW have been reported by Saini et al (2012). In order to find the right biosorbent candidate, it is imperative to screen variety of biomass occurring in human environment.

Free cyanide and thiocyanate, metal-cyanides and metal-thiocyanates can occur in the wastewaters in various forms depending upon the chemical nature of the compounds and the concentration of metal, cyanide and thiocyanate, provided if metal moiety is bound to cyanide and thiocyanate. For example, free cyanide and thiocyanate can occur in the waters in its anionic form like CN^- and SCN^-, respectively. While metal-cyanide like for example - copper-cyanide occur in water in various forms such as $Cu(CN)_2^-$, $Cu(CN)_3^{2-}$, $Cu(CN)_4^{2-}$, etc. Similarly, copper-thiocyanate complex occur in waste waters in various forms like $Cu(SCN)_2^-$, $Cu(SCN)_3^{2-}$, $Cu(SCN)_4^{2-}$, etc. Therefore, it was thought worthwhile to explore the possibility of their recovery by adsorption on low-cost biomass procured from various places. Activated charcoal, a conventional material, was used for obtaining comparative data.

It is well known that the process of biosorption is regulated by aqueous solution pH (Puranik and Paknikar, 1997). The first step in present study was therefore determination of optimum solution pH for biosorption of gold-cyanide and silver-cyanide. It was found that biosorption of both the M_xCNs (for all the low-cost biomass) increased with pH and then declined rapidly with further increase in pH. As seen from the Table 2 that maximum sorption of DCAU was at pH 4.0 while DCAG sorption by most of the biomass was at pH 6.0. Sorption decreased in the alkaline pH. It was found that other than activated charcoal (which was used as reference material) biomass like Rice husk & *Eichhornia* roots and *Eichhornia* roots and Tea powder waste were efficient biosorbents for DCAU and DCAG sorption. There was no auto-oxidation loss of both the M_xCNs in controls without sorbent confirming that (bio)sorption is the only mechanism by which M_xCNs are being removed from solution. In the previous study carried out by Niu and Volesky (2000) found that the maximum adsorption of DCAU by biomass was in the acidic pH ranging from 2.0 to 4.0. These results corroborates with the results obtained in our studies.

Increased sorption under acidic conditions may be due to the protonation of the functional groups acquiring net positive charges. Probably therefore, the formation of species such as H^+-$AuCN_2^-$ and H^+-$AgCN_2^-$ on the biomass might have taken place thereby accommodating more number of M_xCN species on the biosorbent sites. Waste biomass from natural origin contains large number of surface functional groups like hydroxyl, carbonyl, carboxyl, sulphydryl, amine, imine, amide, phosphonate, phosphodiester, etc. Probably some of these functional groups might have played the crucial role in the sorption of DCAU and DCAG from aqueous solution.

Matheickal and Yu (1996) have reported that pH dependence of cationic and anionic adsorption can largely be related to type and ionic state of these functional groups and the chemistry of target compound in solution. DCAU and DCAG in our studies could be compared with anionic metal species like hexavalent chromium (an oxyanion) and arsenic. At low pH values, cell wall ligands are protonated and compete significantly with metal binding. With increasing pH, more ligands such as amino and carboxyl groups, would be exposed leading to attraction between these negative charges and the metals and hence increases in biosorption on to cell surface (Aksu, 2001). As the pH increased further, the overall surface charge on the cells could become negative and biosorption decreased (Aksu, 2001). Patil and Paknikar (1999) have reported the optimum pH of 4.0 for the sorption of copper- and nickel-cyanide from aqueous solutions using *Cladosporium cladospoiroides* biomass.

Free cyanide (CN^-) bearing effluents are highly alkaline in nature and have pH ranging from 9.5 to 12.5, whilst M_xCN effluents have pH in range of 6.0 to 10.0. Obviously, suitable pH alterations of the effluents would be required before biosorption. Unlike free cyanide, M_xCNs does not evolve potent hydrogen cyanide (HCN) gas because of their high stability constants (APHA-AWWA-WEF, 1998;Sharpe, 1976). Therefore, biosorption under acidic conditions would be a safe procedure. On the basis of screening studies under optimum pH conditions, Rice husk and *Eichhornia* root biomass were selected for DCAU sorption, while *Eichhornia* roots and Tea powder waste biomass were selected for DCAG sorption for further experiments.

The DCAU and DCAG loading capacity of the biosorbent could be taken as an equivalent measure of binding sites present. It was found that Rice husk biomass had the maximum loading capacity for DCAU (7.63 µmol/g) sorption among the two tested biomass; and was followed by *Eichhornia* root biomass (7.04 µmol/g). Loading capacity of Activated charcoal was less (7.61 µmol/g) when compared with Rice husk. In case of DCAG biosorption, *Eichhornia* root biomass showed highest loading capacity (9.74 µmol/g) followed by Tea powder waste (9.41 µmol/g). Loading capacity of *Eichhornia*root biomass though marginally less, but was highly competitive and comparable with that of activated charcoal (9.95 µmol/g). This unlocks newer opportunities of developing an efficient biosorption process for the removal and recovery of anionic species like gold-cyanide and silver-cyanide from low tenor waste solutions. In the study carried out by Patil (1999) it was found that the biomass of *C. cladosporoides* had higher loading capacity (34-40 µmol/g) than activated charcoal (27.5-30 µmol/g) for the sorption of metal-cyanides viz. copper- and nickel-cyanide. These results also indicate that more such biomass screening programmes are needed in search of right candidate for efficient sorption.

In the present study loading capacity of conditioned biomass was also compared with that of unconditioned biomass (Table 3 and 4). For DCAU sorption, the unconditioned biomass showed lowered loading capacity compared to conditioned biomass. This reduction in loading capacity might be due to pH at which the loading capacity was determined. For conditioned biomass, the optimum pH for sorption was 4.0 as against the pH of sorption of unconditioned biomass i.e. the pH of original biomass (pH of Rice husk 5.94; pH of *Eichhornia* root biomass 7.01). In case of DCAG biosorption, it was observed that pH of unconditioned and conditioned biomass did not have any effect on the loading capacity of *Eichhornia* root and Tea powder waste biomass. This could be illustrated by the fact that original pH (unconditioned) of both *Eichhornia* root (pH 7.01) and Tea powder waste (pH 5.94) were similar to the obtained optimum pH values of our experiments. This result is very important from the view point of actual use of the biosorption process at commercial scale is concerned. Use of unconditioned biomass at commercial scale will save both time and money thereby making the cost of treatment economical which otherwise would have required for conditioning the biomass. Considering these results, selection of biosorbent was further narrowed down to Rice husk and *Eichhornia* root biomass for DCAU and DCAG biosorption, respectively.

For cost effective treatment of industrial effluents, it is imperative to discern the biomass quantity (i.e. solid-to-liquid ratio) required. In our experiments, it was found that as the biomass quantity increased the % biosorption of both the M_xCNs also increased. Maximum uptake in terms of Q (3.84 µmol/g) was observed at 3% (w/v) of Rice husk biomass for DCAU sorption. However, from 1 to 5 % (w/v) there was no significant increase. In case of DCAG sorption, *Eichhornia* root biomass showed highest Q value for the biomass-to-sorbent quantity from 2.0 to 5.0% (w/v). However, as the concentration of biomass was further increased the M_xCN uptake did not increase the biomass loading which is attributable to the interference between binding sites at higher quantities (de Rome and Gadd, 1987).

Process of biosorption is fundamentally a surface interaction and is characterized by rapid uptake of ions by biomass surfaces. Rapidity of the process makes it a worthy candidate for use in effluent treatment on a commercial scale. Kinetics showed that rate of uptake of both the M_xCN was maximum in first 15-20 minutes with over 80% of biosorption. Later, the sorption rate slowed down until it reached a plateau after 35-40 min, indicating the equilibration of system. Maximum sorption of DCAU and DCAG was 88% and 94% in 40 min. The quick equilibrium time may be attributed to the particle size. The effective surface area is high for small particles. Such type of result is typical for biosorption of metals involving no energy-mediated reactions, where metal removed from solution is due to purely physico-chemical interactions between the biomass and metal in solution. Basha *et al.* (2008) observed similar results in case of biosorption of oxyanion species viz. chromium using seaweed *Cystoseira indica*. The rapid kinetics has significant practical importance as it will facilitate smaller reactor/column volumes ensuring efficiency and economy.

The influence of starting DCAU and DCAG concentration on biosorption by Rice husk and *Eichhornia* roots biomass showed that equilibrium sorption capacity of the sorbent increased with increasing starting concentration of M_xCNs from 0.01 to 1 mM (10 to 1000 µmol).

This might be due to the increase in number of ions competing for available binding sites in the biomass. Uptake of M_xCNs at various concentrations reached a plateau when the concentration was in the range of 0.5 mM (500 μmol). This might be due to the saturation of binding sites, which clearly showed that M_xCN uptake by Rice husk and *Eichhornia* root biomass was a chemically equilibrated and saturable phenomenon. The higher starting concentration of target compound offers increased driving force to overcome all mass transfer resistance of target chemical ions between the aqueous and solid phases resulting in higher probability of collision between M_xCN ions and the biosorbent. This results in higher uptake of the target compound. Moreover, the biomass cell membrane comprises host of functional groups made of polysaccharides, proteins, lipids that have the potential of binding to M_xCN ions.

It is well known that biosorption resembles physical adsorption process and follows an adsorption type isotherm (Tsezos, 1990). Adsorption isotherms are the plots of solute concentration in the adsorbed state as a function of its concentration in the solution at constant temperature. Equilibrium sorption isotherms give useful evidence for selection of an adsorbent and facilitate evaluation of adsorption process for a given application (Weber, 1985). Isotherm indicates the relative affinity of biosorbent for target ions and the adsorption capacity of biosorbent. Also, the sensitivity of biosorption to changes in target compound concentration can be determined by the relative steepness of the isotherm line. Some of the important equilibrium models developed to describe adsorption isotherm relationships include single layer adsorption (Langmuir, 1918; Freundlich, 1926) and multilayer adsorption (Branauer *et al.*, 1938).

Adsorption isotherms are known to have been largely used for projected industrial applications (Tsezos and Volesky, 1981). In the present study, it was decided to fit the DCAU and DCAG sorption data with two most widely accepted adsorption models viz. Freundlich and Langmuir. Linear transformation of the adsorption data using Freundlich and Langmuir models ($R^2 = >0.96$) allowed computation of the M_xCN adsorption capacities. Experimental data was found to obey the basic principles underlying these models, that is, heterogeneous surface adsorption and monolayer adsorption at constant adsorption energy, respectively (Langmuir 1918; Freundlich 1926).

Other than the M_xCN species many additional cations and anions are normally encountered in the effluents emanated from industries like metal mining, electroplating, photofinishing units, printed circuit board manufacturing, etc. These species might inhibit the removal of DCAU and DCAG from aqueous solutions. The impact of commonly occurring cations and anions was therefore studied on biosorption of DCAU and DCAG by Rice husk and *Eichhornia* root, respectively. It was observed that M_xCNs were not significantly affected in most of the cases. However, biosorption of DCAU reduced by 33-40% in the presence of zinc, chromium and cadmium. In case of DCAG, sorption reduced by 37-67% by the presence of cadmium, zinc, iron and chromium. Biosorption in the presence of other metals cations (copper, nickel and silver) and anions (phosphates, sulphates and chlorides) was consistently above 80%.

Pretreated Rice husk and *Eichhornia* biomass with variety of chemicals showed greater degree of variation in the biosorption of DCAU and DCAG. Pretreatment of Rice husk and *Eichhornia* root biomass with 1% L-cysteine enhanced the biosorption capacity of both the M_xCNs, while the NaOH pretreated biomass significantly hampered the biosorption process. Based upon the results obtained, it was thought worthwhile to determine the loading capacity of L-cysteine pretreated biomass as well. It was observed that the loading capacity of Rice husk and *Eichhornia* root biomass enhanced the biosorption of DCAU and DCAG upto 175% and 140%, respectively compared to untreated biomass (i.e. in absence of cysteine loaded biomass). These result corroborated with the findings obtained byNiu and Volesky (2000). This could be explained by the fact that in the acidic pH (pH 4.0 to 6.0 in our study), weak base groups either on cysteine or on the biomass becomes protonated and acquires a net positive charge. Roberts (1992) had reported the pK ranging from 3.5 to 6.0 of the positively charged weak base amine groups. Carboxyl group on the biomass could be protonated in their neutral for as the pKa is in the range of 3 to 5 (Buffle, 1988). In acidic pH range of 2.0 to 6.0, some of the carboxyl groups on cysteine may still be dissociated since the dissociated constant of carboxyl group on cysteine is 1.90, whereas the amino group is protonated and with a positive charge. This allows the cysteine binding to biomass through the integration/combination of negative cysteine carboxyl groups with some of the positively charged biomass functional groups. Thus, the positively charged cysteine amino group were available for binding of anionic M_xCN species like [Au(CN)$_2^-$] and [Ag(CN)$_2^-$] which are the target compounds in our studies. In other words, the anionic species [Au(CN)$_2^-$] and [Ag(CN)$_2^-$] adsorbed by ionizable functional groups on cysteine loaded biomass carrying a positive when protonated.

(Waste Biomass --- Cysteine --- H$^+$) --- Au(CN)$_2^-$

(Waste Biomass --- Cysteine --- H$^+$) --- Ag(CN)$_2^-$

When the target compound is rare and costly, it is always desirable to recover the target compound from industrial effluents having low concentration and high volumes. For an effective and viable biosorption technology, elution methods for the recovery of target compound should be highly efficient, economical and should not cause damage to the biomass. Several eluting agents have been reported in the literature which includes mainly mineral acids, alkalis, organic acids, etc. In the present study, the loaded DCAU and DCAG on Rice husk and *Eichhornia* root biomass, respectively, could be desorbed with more than 95% efficiency using 1 N sodium hydroxide solution. Final concentration of DCAU and DCAG in the concentrated eluent was 28-30 and 22-25 folds, respectively, of the starting concentration. Such high tenor solution of recovered gold-cyanide and silver-cyanide may be recycled back to the user industry.

The next major task in the study was to test the selected biomass viz. Rice husk and *Eichhornia* root biomass for the removal of gold-cyanide and silver-cyanide from their respective industrial effluents in batch mode. As mentioned earlier that the project investigator encountered great difficulty in procuring the effluent samples from industries. In the end, third party intervention helped the investigator to get the sample. In developing country like India, most of industrial personnel are reluctant to give any information regarding toxic chemical waste like cyanide. Moreover, they don't allow the outsider to invade into their industry mainly due to the risk and threat that is associated with cyanide disposal. With the stricter statutory limits imposed by statutory agencies, the conventional physic-chemical methods for the treatment of metal-cyanide bearing effluents are proving to be expensive and also inadequate to meet the required standards. This techno-economic impasse has led to closure of several industries especially the plating industries.

The gold-cyanide and silver-cyanide from the effluents procured from the industries could be effectively biosorbed by Rice husk and *Eichhornia* root biomass which were pretreated with L-cysteine. Gold and cyanide removal efficiency from gold-cyanide industrial effluent was 91.53% and 82.69%, respectively. However, the cyanide level in the biosorbed treated effluent although very less (0.59 mg/l) but was not below the standard limits prescribed by Indian statutory agencies, which is 0.2 mg/l. Similarly, the cyanide concentration after biosorption treatment to silver-cyanide effluent was also not complying with the standards prescribed by Indian statutory agencies. Overall, the studies on industrial effluents indicated that both the biomass viz. Rice husk and *Eichhornia* root biomass were very effective in treating both the effluents by biosorption process. Therefore, it is possible to employ Rice husk and *Eichhornia* root biomass for the treatment of industrial effluents on commercial scale. The residual (unrecoverable) cyanide remaining in the solutions after biosorption were subjected to biodegradation process using bacterial consortium under optimized conditions.

When the residual gold-cyanide and silver-cyanide biodegradation experiment was run under optimized conditions in batch mode, it was found that the live bacterial consortium previously isolated by Patil (2008) could degrade the cyanide present in the solution within a period of 5 h with an efficiency of >90% for both types of effluents. The resulting treated solution could comply with the disposal standards prescribed by statutory agencies in India. These findings indicated that biodegradation could be used as a polishing step in the treatment of precious M_xCNs containing industrial waste waters.

In process applications, the most effective apparatus for sorption/desorption and making the most effective use of the reactor volume, is a fixed-bed column. The column makes optimum use of the concentration gradient between the solute sorbed by the solids and that remaining in the liquid phase thereby providing the driving force for the biosorption process. The process of biosorption (metals and their related species) is governed by three key regimes: (i) the sorption equilibrium, (ii) the sorption particle mass transfer and (iii) the flow pattern through the packed bed. These three regimes determine the overall performance of the sorption column which is judged by its 'service time'. Service time is the length of time until the sorbed species breaks through the bed to be detected at a given concentration in the column effluent. The breakthrough point indicates that the column is saturated practically and could be taken out of operation for some kind of its regeneration (Volesky, 2003).

The column bed is being saturated at inflow concentration which represents equilibrium concentration for the part of the bed upstream from the transfer zone. The saturation of the bed/column varies from zero to the full saturation. This zone of partial saturation moves the column in the direction of the flow at a certain velocity which is predominantly determined by the biomass loading, sorbent capacity and the feed rate to column. The column is operational until this zone reaches the end of the column. Until that time the effluent leaving the column has no trace of the sorbate in it. When the transfer zone reaches the column end, the sorbate concentration in the effluent starts to gradually increase and for all practical purposes, the working life of the column is over and the "breakthrough point" occurs marking the usable column "service time". These two parameters are very important from the process design point of view because they directly affect the feasibility and economics of the sorption process (Volesky, 2003).

After successfully treating both the industrial effluents in batch mode using Rice husk and *Eichhornia*root biomass, further biosorption studies were carried out in continuous mode using packed bed column. It was found that the service time offered by the column beds for gold-cyanide (from column 1) and silver-cyanide (from column 2) effluents were 60 h and 40 h, respectively. In other words, these itself were the breakthrough points. For both the columns the transfer zone observed was of 30 h each. The total effluent passed through the column 1 and 2 was equivalent to 50 and 34 bed volumes, respectively, while the complete saturation occurred after 90 and 70 h, respectively. Continuous study clearly showed that both the effluents were biosorbed and treated successfully in the packed bed columns for the removal of both precious and toxic species. Further, in these studies the project investigator did not immobilized any of the biomass primarily because the present work was focused on low tenor effluents containing precious gold and silver and toxic chemical species like cyanide (all below 10 mg/l). Secondly, the results obtained through batch and continuous studies showed that both the biomass were efficient enough to sorb and treat the effluents and therefore the project investigator felt that immobilization of the biomass probably is not required in this case.

Thus, it could be concluded that the waste biomass used in the present study has immense potential "as biosorbents" for the removal/ management of low tenor precious and toxic pollutants, as evident from the example of gold-cyanide and silver-cyanide management in the present study. Further, biosorption technology used could also become an economical, non-destructive and reliable alternative to the conventional processes for the management of industrial effluents employed on the commercial scale.

APPLICATION OF BIOSORPTION TO SOME NEWER WASTES AND PRODUCTS

Apart from the removal and recovery of precious heavy metal species from industrial effluents, passive bioremediation technology (PBT) can also be employed for some newer type of wastes and products that have emerged in the recent times.

A novel approach of combined biosorption-biodegradation processes was used by Patil and Paknikar (1999) for the removal and recovery of copper- and nickel-cyanide from electroplating effluents. *Cladosporium cladosporioides* biomass was found to be highly efficient sorbent in this case. The unrecoverable (residual) metal-cyanides after biosorption was subjected to biodegradation process using bacterial consortium. The treated effluent was free of cyanide and metals and complied with the statutory limits (Patil and Paknikar, 1999).

The problem of waste photovoltaic cells was addressed by Paknikar et al (1997) by way of recovering and recycling of expensive metals like silver, cadmium and tellurium. In this study, the researchers used scrapings of waste photovoltaic cells, which were dissolved in suitable mineral acid and was diluted to obtain desired metal concentration. The metal solution was then subjected to biosorption column consisting of inactive granulated biomass of *C. cladosporioides* #1 for selective removal of silver. Similarly, in the next step, cadmium was removed by biosorption process by passing the solution through biosorbent column.

The silver and cadmium free solution after treatment was then fed to tellurium reducing bioreactor consisting *P. mendocina*. The overall removal and recovery efficiency of these metals was >90%. With the rising demand and shift towards renewable energy sources, the number of photovoltaic cells producing units/industries will increase in the years to come; and so the use of non-renewable resources like metals and it wastages in the form of rejections. Although the economic feasibility of the process was not studied by the researchers) but the study certainly add to the advancement of knowledge by employing combined passive and active bioremediaton (Paknikar et al 1997).

Pethkar et al. (2001) reported an interesting study on the removal of toxic metals like lead and cadmium from fruit juices of carrot, grapes and oranges, and extracts of Jatamansi herb and raisin by passive bioremediation using the biomass of *C. cladosporioides* #2. With a growth rate of 15%, the annual turnover of herbal medicinal industry in India is Rs. 75,000 million. As per ASSOCHAM (Associated Chamber of Commerce and Industry), the turnover of herbal industry is projected to double to Rs.1,50,000 million (USD 3 billion) by 2015. However, the business is getting severely affected by the presence of toxic heavy metals into food and herbal products thereby making them unacceptable in foreign markets because of their stringent statutory norms. Therefore, removal of these toxic metals from such products using biosorption process is crucial and has great prospectus. Sun et al (2007) had reported sorption of heavy metal ions by polyaspartyl polymers from Chinese herbal medicines. However, there is paucity of literature on biosorption of toxic metals from herbal medicines and food products.

Bhat et al (2012) had proposed a novel integrated model for the recovery of gold/silver from e-waste using an integrated hydrometallurgical (chemical) and biometallurgical (low cost biomass) processes. Feasibility study was conducted to explore the possibility of removal/recovery of silver-cyanide using low-cost biosorbents. *Eicchornia* root biomass and Waste tea powder were found to be an efficient low-cost biosorbents for leached silver-cyanide from electronic scrap. The concentrated silver-cyanide recovered had the potential for its further use as input material for electroplating industry (Bhat et al, 2012). Awareness among the urban population regarding disposal and management e-waste has also been studied by Bhat and Patil (2012).

In the twenty-first century, entire world is witnessing a paradigm shift in the overall waste management practices, which is rapidly changing its face and orientation. Waste is no more considered as waste but is recognized as a 'Resource'. This lost resource could potentially be recovered from the wastes using suitable strategies and technologies. Therefore, in a real sense, model like recovery and recycling of waste resource is gaining remarkable importance in today's so called 'Technological Society'. Application of concepts similar to this work will ultimately reduce the demand for natural resources thereby extending its sustenance. In view of this, the present chapter on passive bioremediation will certainly add to the advancement of knowledge in the field of resource recovery and industrial pollution management, waste minimization and will help profitability of business community at large. It has not escaped through authors mind that the recovered resource from the waste of one industry has all the potential for its use as an input material for other industry thereby strengthening the emerging discipline of 'Industrial Ecology'.

ACKNOWLEDGEMENTS

Dr. Yogesh Patil gratefully acknowledges the International Foundation for Science (IFS), Stockholm, Sweden, in cooperation with The Organization for the Prohibition of Chemical Weapons (OPCW), The Hague, The Netherlands, for providing the research grant.

REFERENCES

1. Aksu Z (2001) Equilibrium and kinetics modeling of cadmium (II) biosorption by *C. vulgaris* in a batch system: effect of temperature. *Separation and Purification Technology* 21:285-294.

2. APHA, AWWA, WEF (1998) Standard Methods for the Examination of Water and Wastewater Analysis, 20th Ed. American Public Health Association, Washington, DC.

3. Azab HM, El-Shora HM and Mohammed HA (1995) Biosorption of cyanide from industrial waste water. *Al-Azhar Bulletin of Science* 6:311-323.

4.   Basha S, Murthy ZVP and Jha B (2008) Biosorption of hexavalent chromium by chemically modified seaweed, *Cystoseira indica*. *Chemical Engineering Journal* 137:480-488.

5.   Bhat Viraja and Patil Y.B. (2012) Mobile user's perspective towards e-waste: A case study of Pune city. *International Journal of Academic Conference Proceedings* 1(2).

6.   Bhat Viraja, Rao Prakash and Patil Yogesh (2012) Development of an integrated model to recover precious metals from electronic scrap - A novel strategy for e-waste management. *Procedia - Social and Behavioral Sciences* 37: 397-406.

7.   Braunauer S *et al* (1938) Adsorption of gases in multimolecular layers. *Journal of American Chemical Society* 60: 309-319.

8.   DeRome L and Gadd GM (1987), Copper adsorption by Rhizopus arrhizus, Cladosporium resinae and Penicillium italicum. *Applied Microbiology and Biotechnology* 26: 84-90.

9.   Eckenfelder WW (1989) Industiral Water Pollution Control. New York, McGraw Hill.

10.   Freundlich H (1926) *Colloid and Capillary Chemistry,* London: Methuen.

11.   Gaddi Shivanand S and Patil Yogesh B (2011) Screening of some low-cost waste biomaterials for the sorption of silver-cyanide [Ag(CN)2-] from aqueous solutions. *International Journal of Chemical Sciences* 9: 1063-1072.

12.   Ganczarczyk JJ, Takoaka PT, Ohashi DA (1985) Application of polysulfide for pretratment of spent cyanide liquors. *Journal of Water Pollution Control Federation*, Cont. Fed., 57: 1089-93.

13.   Hasany SM, Saeed MM and Ahmed M (2001) Sorption of traces of silver ions onto polyurethane foam from acidic solution. *Talanta* 30(1): 89-98.

14.   Itankar, Nilisha and Patil, Yogesh (2012) Sorption of toxic hexavalent chromium (Cr6+) from aqueous solutions by some (bio)materials. In:*Proceedings of International Conference on Functional Materials for Defence (ICFMD)* jointly organized by Deference Institute of Advanced Technology, Pune, Naval Postgraduate School, USA & Office of Naval Research (Global), USA, during 18-20 May 2012, p. 137.

15. Karavaiko GI, Kondrat'eva TF, Savari EE, Grigor'eva NV and Avakyan ZA (2000) Microbial degradation of cyanide and thiocyanate *Microbiology*69(2): 167-173.

16. Kiruthika AJ and Shrinithya (2008) Cyanide detoxification and recovery of gold from gold effluent. *Advanced Biotech* pp. 20-26.

17. Kononova ON, Kholmogorov AG, Danilenko SV, Kachin SV, KononovYS and Dmitrieva, ZV (2007) Sorption of gold and silver on carbon adsorbents from thiocyanate solutions. *Carbon* 43(1): 17-22.

18. Langmuir J (1918) The adsorption of gases on plane surfaces of glass, mica and platinum. *Journal of American Chemical Society* 40: 1361-1403.

19. Lee JS, Deorkar NV & Tavlarides LL (1998) Adsorption of copper cyanide on chemically active adsorbents. *Industrial and Engineering Chemical Research.*, 37: 2812-2820.

20. Matheickal JT and Yu Q (1996) Biosorption of lead from aqueous solutions by marine alga *Ecklonia radiate*. *Water Science and Technology* 34: 1-7.

21. Modak JM and Natarajan KA (1995) Biosorption of metals using nonliving biomass – A review.*Mineral and Metallurgy Process*12: 189-196.

22. Mohan D and Pittman CU Jr. (2006) Activated carbons and low cost adsorbents for remediation of tri- and hexavalent chromium from water.*Journal of Hazardous Materials*, B137: 762-811.

23. Namasivayam C and Sangeetha D (2007) Kinetic studies of adsorption of thiocyanate onto zinc chloride activated carbon from coir pith, an agricultural solid waste. *Chemosphere* 60(11): 1616-1623.

24. Niu H & Volesky B (2001) Gold adsorption from cyanide solution by chitinous materials. *Journal of Chemical Technology and Biotechnology*, 2001, 76(3), 291-297.

25. Niu H and Volesky B (2000) Gold-cyanide biosorption with L-cysteine. *Journal of Chemical Technology and Biotechnology* 75: 436-442.

26. Paknikar et al (1997) An integrated chemical-microbiological approach for the disposal of waste thin film cadmium telluride

photovoltaic modules. In: *Environmental, Safety and Health Issues in IC Production,* edited by R Reif et al. Materials Research Society, Pittsburgh, Pennsylvania, USA, pp. 133-138.

27.   Paknikar KM, Pethkar AV & Puranik PR (2003) Bioremediation of metalliferous wastes and products using inactivated microbial biomass. *Indian Journal of Biotechnology.* 2: 426-443.

28.   Patil Yogesh B (1999) Studies on biological detoxification of metal-cyanides containing industrial effluents. Ph.D. Thesis, University of Pune, Pune, India.

29.   Patil YB and Paknikar KM (1999a) Removal and recovery of metal cyanides using a combination of biosorption and biodegradation processes.*Biotechnology Letters* 21: 913-919.

30.   Patil YB and Paknikar KM (1999b) Removal and recovery of metal cyanides from industrial effluents. *Process Metallurgy* 9: 707-716.

31.   Patil YB and Paknikar KM (2000a) Biodetoxification of silver-cyanide from electroplating industry wastewater. *Letters in Applied Microbiology*30: 33-37.

32.   Patil YB and Paknikar KM (2000b) Development of a process for biodetoxification of metal cyanides from wastewaters. *Process Biochemistry* 35: 1139-1151.

33.   Patil YB and Paknikar KM (2001) Biological detoxification of nickel-cyanide from industrial effluents. *Process Metallurgy* 11: 391-400.

34.   Patil Yogesh B (2006) Isolation of thiocyanate degrading chemoheterotrophic bacterial consortium. *Nature Environment and Pollution Technology* 5(1): 135-138.

35.   Patil Yogesh B (2008a) Biodegradation of thiocyanate from aqueous waste by a mixed bacterial community. *Research Journal of Chemistry and Environment* 12(1): 69-75.

36.   Patil Yogesh B (2008b) Thiocyanate degradation by pure and mixed bacterial cultures. *Bioinfolet* 5(3): 308-309.

37.   Patil Yogesh B (2011) Utilization of thiocyanate (SCN-) by a metabolically active bacterial consortium as the sole source of nitrogen. *International Journal of Chemical, Environmental & Pharmaceutical Research,* 2: 44-48.

38.   Patil Yogesh B (2012) Development of a low-cost industrial waste treatment technology for resource conservation – An urban case

study with gold-cyanide emanated from SMEs. *Procedia – Social and Behavioural Sciences* 37: 379-388.

39. Patil Yogesh B and Kulkarni Anil R (2008) Environmental sensitivity and management of toxic chemical waste in mining industry with special reference to cyanide. In: *High Performing Organizations: Needs and Challenges*, Tata McGraw Hill Publications, Part D: pp. 183-196.

40. Patil Yogesh, ChoureyJayati and RaoPrakash (2012) Biotechnological strategy for the management of industrial waste with concurrent mitigation of global warming – A feasibility study using microalgae. In: *Inclusiveness and Innovation – Challenges for Sustainable Growth of Emerging Economies* (Edited by Rajiv Divekar and BR Londhe), Excel India Publishers, New Delhi, India, pp. 249-255.

41. Pethkar AV et al. (2001) Biosorptive removal of contaminating heavy metals from plant extracts of medicinal value. *Current Science* 80, 1216-1219.

42. Prashanth MS and Patil Yogesh B (2007) Behavioural surveillance of Indian major carp *Catla catla* (Hamilton) exposed to free cyanide. In:*Disaster Ecology and Environment*, Book Edited by Arvind Kumar, Daya Publishing House, New Delhi, pp. 210-215.

43. Roberts GAF (1992) Chitin Chemistry, Macmillan, London, UK, pp. 204-206.

44. Rollinson G, Jones R, Meadows MP, Harris RE and Knowles CJ (1987) The growth of a cyanide-utilizing strain of *Pseudomonas fluorescens* in liquid culture on nickel cyanide as a source of nitrogen *FEMS Microbiology Letters* 40: 199-205.

45. Saini Samir, RaoPrakash and Patil Yogesh (2012) City based analysis of MSW to energy generation in India, calculation of state-wise potential and tariff comparison with EU. *Procedia - Social and Behavioral Sciences* 37: 407-416.

46. Sharpe AG (1976) The chemistry of cyano complexes and the transition metals. London: Academic Press, 1976.

47. Sun Bo, Mi Zhen-Tao, An Gang and Liu Guozhu (2007) Binding of several heavy metal ions by polyaspartyl polymers and their application to some Chinese herbal medicines. Journal of Applied Polymer Science. 106(4): 2736-2745.

48. Thakur Ravindra Y and Patil Yogesh B (2009) Management of thiocyanate pollution using a novel low cost natural waste biomass. *South Asian Journal of Management Research* 1(2): 85-96.

49. Tsezos M (1990) Engineering aspects of metal bonding by biomass. In: *Microbial Mineral Recovery*, edited by HL Ehrlich and CL Brierley, McGraw-Hill Publishing Co., New York, USA, pp. 325-339.

50. Tsezos M and Volesky B (1981) Biosorption of uranium and thorium. *Biotechnology and Bioengineering* 23: 583-604.

51. Vapur H, Bayat O, Mordogan H and Poole C (2005) Effects of stripping parameters on cyanide recovery in silver leaching operations.*Hydrometallurgy* 77(3-4): 279-286.

52. Volesky B (2003) *Sorption and Biosorption*, BV Sorbex, Inc, Montreal – St. Lambert, Quebec, Canada.

53. Weber WJ Jr. (1985) Adsorption theory, concepts and models. In: *Adsorption Technology: A Step-By-Step Approach to Process Evaluation and Application*, Edited by FL Slejko, Marcel Dekker, New York, pp. 1-35.

Preparation and Stability of Inorganic Solidified Foam for Preventing Coal Fires

Botao Qin[1,2], Yi Lu[2], Fanglei Li[2], Yuwei Jia[2], Chao Zhu[2], and Quanlin Shi[2]

[1]State Key Laboratory of Coal Resources and Mine Safety, CUMT, Xuzhou, Jiangsu 221008, China

[2]Faculty of Safety Engineering, China University of Mining and Technology, Xuzhou, Jiangsu 221116, China

ABSTRACT

Inorganic solidified foam (ISF) is a novel material for preventing coal fires. This paper presents the preparation process and working principle of main installations. Besides, aqueous foam with expansion ratio of 28 and 30 min drainage rate of 13% was prepared. Stability of foam fluid was studied in terms of stability coefficient, by varying water-slurry ratio, fly ash replacement ratio of cement, and aqueous foam volume alternatively. Light microscope was utilized to analyze the dynamic change of bubble wall of foam fluid and stability principle was proposed. In order to further enhance the stability of ISF, different dosage of calcium fluoroaluminate was added to ISF specimens whose stability coefficient was tested and change of hydration products was detected by scanning electron microscope (SEM). The outcomes indicated that calcium fluoroaluminate could enhance the stability coefficient of ISF and compact hydration products formed in cell wall of ISF; naturally, the stability principle of ISF was proved right. Based on above-mentioned experimental contents, ISF with stability coefficient of 95% and foam expansion ratio of 5 was prepared, which could sufficiently satisfy field process requirements on plugging air leakage and thermal insulation.

INTRODUCTION

Coal fires are difficult, persistent, and costly problems worldwide in coal mining processes [1, 2]. They lead to serious environmental issues, safety problems, and considerable economic losses [3]. Meanwhile, spontaneous combustion of coal and heat transfer occurs more frequently due to subsidence and increased channels of air leakage in the goaf. Air leakage prevention and thermal insulation can lower effectively the spontaneous combustion risk of coal [4, 5]. Based on the various techniques for control and extinguishment of coal fires developed and applied over past 60 years [6–8], materials with pore structure are drawing growing attention because of their characteristics involving heat insulation, fire resistance, lightweight, superior fluidity, environmental friendliness, and so on [9, 10].

In this work, a novel material, ISF, with high closed porosity and uniform pore distribution was prepared via mixing aqueous foam and composite slurry consisting of fly ash, cement, and compound additives. In this process of preparation, the following two points are worthy of consideration. Firstly, stable aqueous foam is required for ISF to plug air leakage in mining applications. Furthermore, the stability of foam may be affected by foam generator parameters, surfactants, and their concentration [11]. Selection of surfactants has an impact on the properties of foam as it affects the surface tension and gas-liquid interfacial properties. Secondly, the foam slurry after mixing is a three-phase system (gas-liquid-solid), which should be uniform and stable. But few scholars have studied the foam formation and stabilization in this kind of system. Current researches mainly focus on the stabilization of two-phase foam (aqueous foam or other liquid foam) [12]. Just a few scholars such as Gonzenbach et al. [13], Hunter et al. [14], Sethumadhavan et al. [15], and Vijayaraghavan et al. [16] carried out experimental studies and mechanism analyses on solid particles stabilized aqueous foam.

In this paper, as a first step, we studied the preparation process of ISF and analyzed the working principles and the effects of key devices. As a next step, the aqueous foam with low drainage rate and high expansion ratio was prepared based on sodium dodecyl sulfate (SDS) solution and modified by the foam stabilizers such as cetrimonium bromide (CTAB), sodium chloride (NaCl), and lauryl alcohol (LA). Then the factors influencing the stability coefficient and foam expansion ratio of ISF were investigated. At last, through the observation on drainage of the bubble wall and the hydration products accelerated by calcium fluoroaluminate, the stabilization mechanism of foam fluid was proposed.

EXPERIMENTAL

Raw Materials

Constituent materials are listed below.

- Portland cement (PC) with the compressive strength of 64.5 MPa at 28 days, conforming to BSEN 197-1 type I cement [17].
- Fly ash (FA) with a median particle size of 35 μm and loss on ignition (LOI) of 5.0%, conforming to BS EN 450 [18].
- Calcium fluoroaluminate ($11CaO \cdot 7Al_2O_3 \cdot CaF_2$): it influences the rate of cement hydration, leading to a reduction in setting time.
- Redispersible polymer powder (PP): it is a kind of polymeric powder which can be easily reemulsified in water to reform liquid emulsion with essentially identical properties to the original emulsion.
- Water (W): its percentage was fixed in order to satisfy both the workability criterion and the controlled low strength materials (CLSM) recommendations for the insulation materials [19].

- Lauryl sodium sulfate (SDS), Cetrimonium Bromide (CTAB), NaCl, and lauryl alcohol (LA). They were diluted with water in different ratios.

Preparation Process of ISF

The basic preparation process can be divided into three parts including mixing the composite slurry, preparing aqueous foam, and mixing composite slurry, accelerator, and aqueous foam. We admixed cement, fly ash, and redispersible polymer powder together and got the blend of these three basic raw materials. Then, part of water was injected into the blend and composite slurry formed under the work of stirrer. The water-solid ratio was controlled slightly less than preset ratio. At the same time, the rest of the water was used to dilute the surfactant. Then, high pressure air was pumped into the foam generator and aqueous foam was produced. The next procedure was to mix composite slurry with aqueous foam in a self-made mixer, with some compound additives added. At last, foam fluid was produced and evolved into ISF at room temperature. The specific preparation procedures of ISF are schematically shown in Figure 1. The main installations are shown in Figure2.

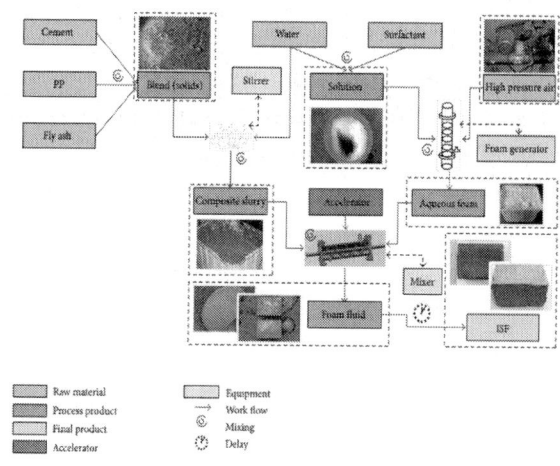

Figure 1: The schematic of the preparation of ISF.

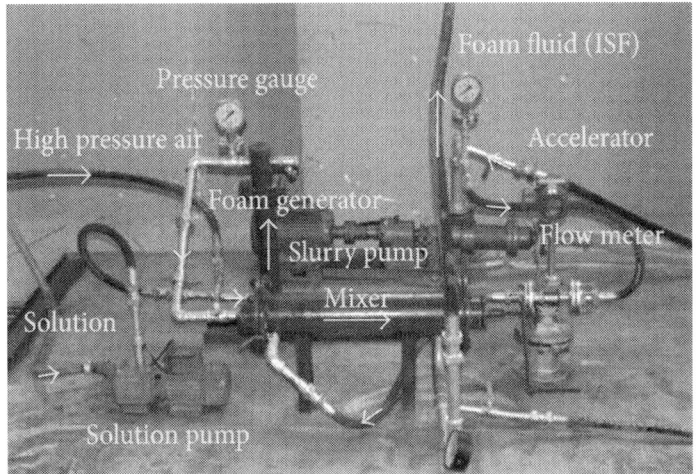

Figure 2: The main installations.

Test Procedure

Drainage Rate Test

We chose the drainage rate to be 30 minutes since aqueous foam was produced, to reflect foam stability. After generation of aqueous foam, the initial foam mass, m, was measured immediately and then poured fully into a Buchner funnel. A measuring cylinder was placed under the Buchner funnel, and the mass of liquid drained from aqueous foam, m_d, was calculated every ten minutes, until 30 minutes. Drainage rate, d_t, can be expressed by

$$d_t = \frac{m_d}{m} \times 100\%.$$

(1)

Stability Coefficient Test

The test instrument was one cylindrical gauge whose inner diameter was 100 mm and measuring range was 315 mm. The test procedure is as follows: (i) pour the fresh ISF into the test instrument and record the initial height, h_0; (ii) measure the final height (h_i) when ISF turns into solidification state. The stability coefficient, ψ, can be calculated according to the following:

$$\psi = \frac{h_i}{h_0} \times 100\%.$$

(2)

Foam Expansion Ratio Test

Test procedure for foam expansion ratio of aqueous foam or foam fluid is as follows: (i) fill a container (volume and mass are known and designated by V_c, m_c, resp.) with surfactant solution or cement slurry, weigh the total mass, m_L, and calculate the density of surfactant solution or cement slurry, ρ_L, by (3); (ii) overfill the aforementioned container with aqueous foam or foam fluid and strike off the excess foam, weigh the total mass, m_F, and calculate the density of aqueous foam or foam fluid, ρ_F, by (4); (iii) calculate the foam expansion ratio of aqueous foam or foam fluid, E_F, by (5) as follows:

$$\rho_L = \frac{m_L - m_c}{V_c},$$

(3)

$$\rho_F = \frac{m_F - m_c}{V_c},$$

(4)

$$E_F = \frac{\rho_L}{\rho_F}.$$

(5)

Microscopy Observation

The microstructure of aqueous foam and ISF fluid were observed by a Nomarski-type phase contrast interference microscope equipped with a digital camera, which can be used to take photomicrograph of the samples and the foams. A drop of sample was brought onto a microscope slide and the structure of bubble was observed. The bubble wall of ISF was investigated by scanning electron microscopy (SEM) (FEI QuantaTM 250 SEM system) with the size of test specimen being $10 \times 10 \times 10\,mm^3$ prism.

Particle Contact Angle Measurement

The water and surfactant solution contact angle on the particles was measured using the gravimetric version of the Washburn method. The method is based on measuring the penetration rate of a wetting liquid into a packed bed of particles, which lead to the following equation [20]:

$$M^2 = t \times \frac{\gamma_{LG}\delta^2 \cos\theta}{\mu} \times \frac{rS^2\varepsilon^2}{2},$$

(6)

Where M is the measured mass of the penetrated liquid; t is the penetration time; γ_{LG} is the gas-liquid surface tension; δ is the liquid density; S is the cross-sectional area of the tube; ε is the void fraction of particles; μ is the viscosity of liquid; r is the mean radius; θ is the contact angle.

Test Design

In order to investigate the influencing mechanism of aqueous foam volume (FV), fly ash replacement for cement (FA), and water-solid ratio (W/S) on the stability of ISF, we conducted tests on different specimens. FV was controlled to vary from 2 V to 10 V with the increment being 2 V, FA changed as 10%, 20%...50%, and W/S increased from 0.3 to 0.5 with every difference quantity being 0.05. According to this design, we conducted 125 tests.

RESULTS AND DISCUSSION

The Key Devices

To develop fine, uniform, and stable inorganic solidified foam, the following two points deserve consideration. Firstly, the foam generator should be able to produce aqueous foam with uniform pore structure, high expansion ratio, and a certain stabilization time. Secondly, aqueous foam and composite slurry should contact thoroughly and then form stable foam fluid during the mixing process in the mixer. The schematic of key devices was shown in Figure 3.

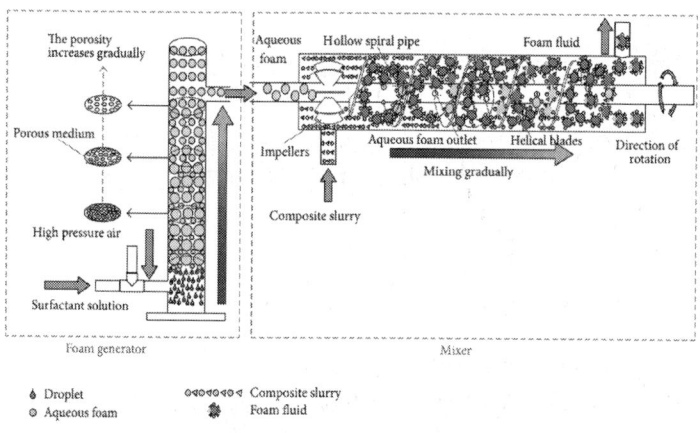

Figure 3: The schematic of foam generator and self-made mixer.

The main process of generating foams by the home-made foam generator is as follows: once foaming agent solution and high pressure air flow through the T-shape conduit of foam generator, the turbulent eddy is formed after mixing and enhanced by the porous medium which can be composed of multilayer meshes, powdered metal, or spherical glass particles, causing greater pressure drop due to their impediment. The more homogenous and denser aqueous foam is produced from down to up as the porosity of porous medium increases stepwise. The aqueous foam produced by mechanical agitation and home-made foam generator was as shown in Figure 4.

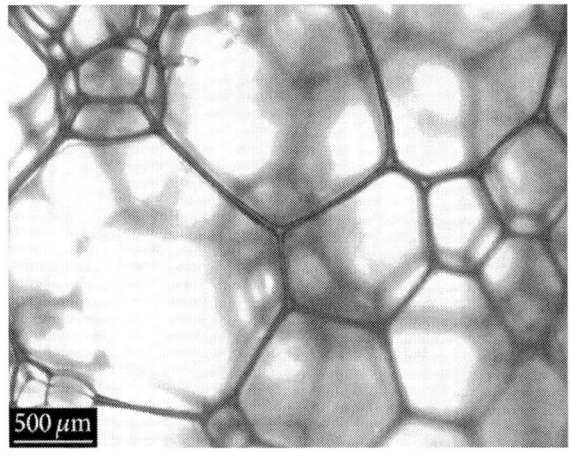

(a)

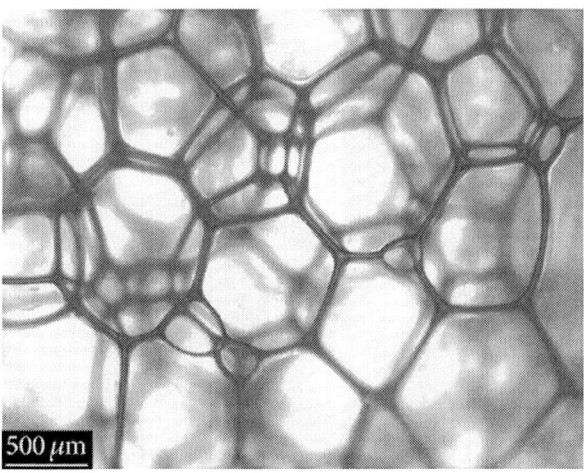

(b)

Figure 4: The optical microscopic analysis diagram of aqueous foam.

Mixer consists of chamber and hollow spiral pipe inside it. The high-speed composite slurry drives the impellers to rotate, and then foam slurry is stirred and delivered by hollow spiral pipe with helical blades. Vortex streets in this process can completely go into turbulence and cause vortex according to certain frequency. The loss of kinetic energy acts on the mixtures and a large number of foam fluids are formed. Aqueous foams pass into the mixer from the left body of hollow spiral pipe equipped with five aqueous foam outlets with an interval angle. Aqueous foams are added to slurry step by step, which reduce the broken rate of foam and increase foam slurry contact areas. This kind of mixing chamber can weaken the shock caused by larger flow of aqueous foam and is conducive for gas-liquid-solid to mix thoroughly.

Preparation of Aqueous Foam

From viewing of the technology process for preparing the ISF, the stability is mainly dependent on that of aqueous foam. Generally speaking, foam expansion ratio of aqueous foam should be more than 20. SDS is a widely used surfactant with strong foaming ability. Its change trends of 30 min drainage rate and foam expansion ratio with different SDS concentrations are depicted in Figure 5.

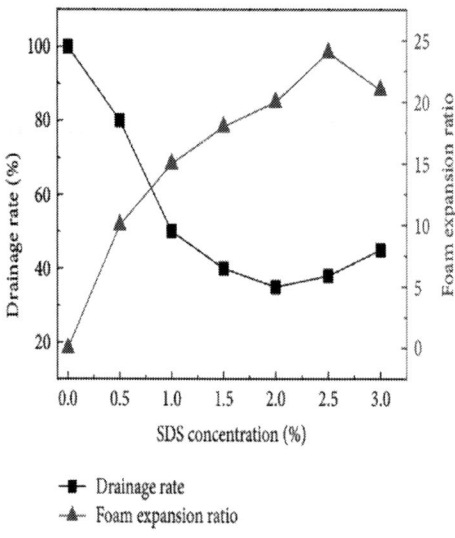

Figure 5: Change trends of drainage rate and foam expansion ratio with different concentrations.

From Figure 5, with increasing concentration of SDS, foam expansion ratio increases firstly and then decreases, for the reason that the surface tension of surfactant solution decreased firstly and then increased due to formation of surfactant micelle, and the largest foam expansion ratio is 24 under a concentration of 2.5%. Drainage rate of aqueous foam presented a reverse trend compared to that of foam expansion ratio, whose minimum is 35% under a concentration of 2%. This is for the reason that more micelles formed and their shape changed with the increase of SDS, contributing to more stable foam films and less drainage rate. However, in the other limit, that is, above 2%, the violation of the law at higher micelle concentrations is related to the appearance of a freezing transition in foam films [21]. Considering the above two indexes, the optimal SDS concentration is 2.5%.

In order to strengthen stability of aqueous foam, CTAB, NaCl, and LA were utilized as foam stabilizers. We studied modification effects on SDS aqueous foam under different concentrations of foam stabilizers ranging from 0.5% to 4.0%, whose concrete effects on foam expansion ratio and drainage rate are shown in Figure 6.

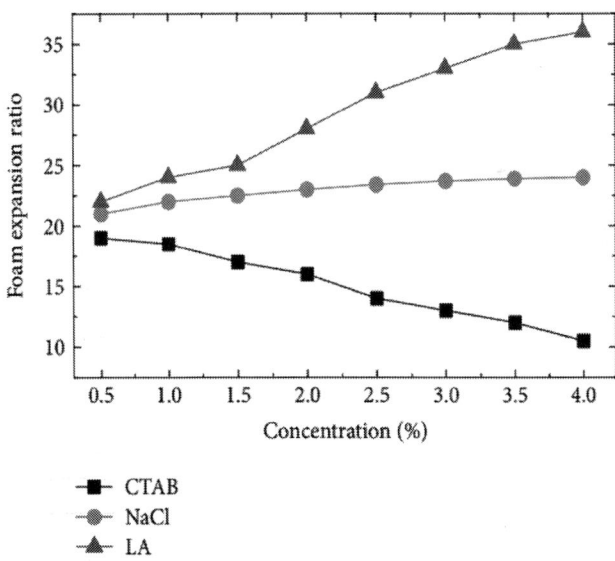

(a)

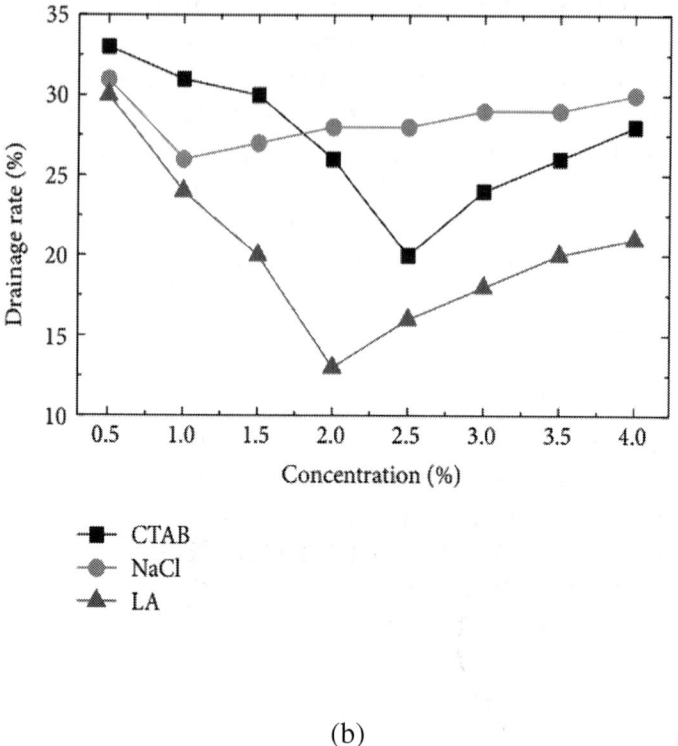

(b)

Figure 6: Modifying effects of foam stabilizer on drainage rate and foam expansion ratio.

From Figure 6(a), the change trends of foam expansion ratio for three foam stabilizers are different, and with increasing concentration, that of CTAB declines and NaCl increases slightly, while LA elevates. In Figure 6(b), from the viewpoint of drainage rate, three foam stabilizers wholly could diminish the drainage rate of aqueous foam, specifically, with the increase of concentration; the drainage rate firstly falls off sharply and tardily goes up later. The minimums of drainage rate and the critical concentrations for CTAB, NaCl, and LA are (20%, 2.5%), (26%, 1.0%), and (13%, 2.0%), respectively. The reasons accounting for the trends mentioned above are special as follows.

Under the condition that the concentration of SDS is 2.5%, its foam expansion ratio decreases with the increasing concentration of CTAB. Because CTAB is a cationic surfactant while SDS is an anionic one, when these two surfactants are mixed, phase separation will occur due to intense electrostatic interaction and condensation of surfactant molecules [22], followed by the ascent of surface tension. With the increase of CTAB concentration, the drainage rate of aqueous foam decreases firstly and then increases, which is 20% and the least under a concentration of 2.5%. Compared with the individual SDS system, the SDS+CTAB mixed system had a synergic effect on foam stabilization [23]. Surfactant mixtures could create a mixed surfactant layer at gas/liquid interfaces. When two bubbles are approaching each other to form a thin liquid film, this mixed surfactant layer can confer disjoining pressures to hinder this approaching.

The foam expansion ratio enlarges with the increase of NaCl concentration mainly because homo-ion could not only diminish the Critical Micelle Concentration (CMC) of the surfactant but also reduce surface tension of the solution and develop its foaming ability. The drainage rate decreases firstly and then ascends with the increase of NaCl concentration, the minimum of which is 26% at a concentration of 1.0%. The addition of NaCl to SDS solution enlarged its foaming ability to some degree and reduced its drainage rate, which could be explained that there is a threshold of added electrolyte on stratification phenomenon of foam film, above which the phenomenon is not observed [24]. Based on our experimental results, we believe that 1.0% was just the threshold. Above 1.0% concentrations of NaCl, bubbles ruptured asynchronously owing to different surface concentrations of NaCl; thus the drainage rate of foam rose slightly with the increased concentrations of NaCl.

The addition of LA could both prominently improve the foam expansion and greatly enhance the stability of aqueous foam. This is because the iceberg structure (a perfectly ordered structure formed by the LA molecules and water molecules) around the hydrocarbon chain in the alcohol makes it a spontaneous process for the alcohol to participate in the formation of micelle and thus bubble films are consolidated. The drainage rate of foam film will slow down with the rise of surfactant micelles in certain range of concentrations [25].

The prepared foam-forming solution containing SDS concentration of 2.5% and LA concentration of 2% possesses excellent foam expansion ratio with the value being 28 and the aqueous foam derived from the solution acquires the best stability with the value being 13%.

The Stability of ISF

Aqueous Foam Volume, Fly Ash Replacement for Cement and Water-Solid Ratio

According to the results of 125 tests, it can be concluded that when FV is 8 V, FA is 30% and W/S is 0.4, and the ISF is in the best state with its foam expansion ratio and stability coefficient being 5 V and 90%, respectively. At the same time, some other test data was shown in Figure 7.

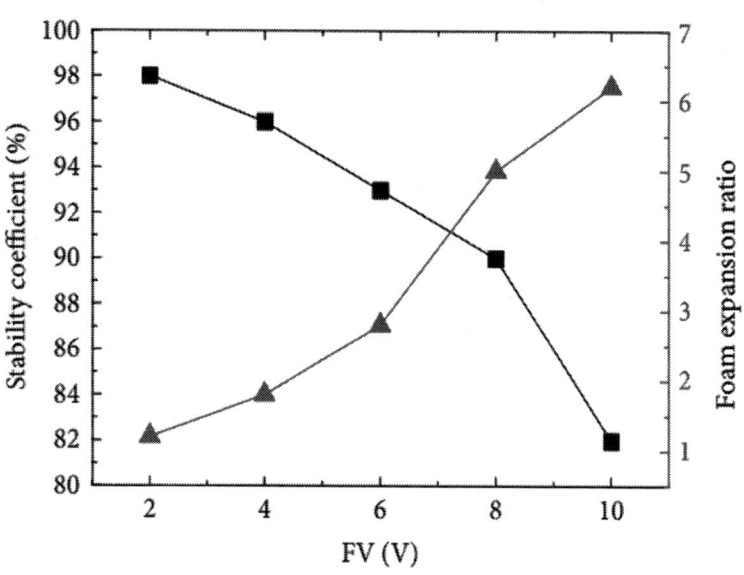

(a)

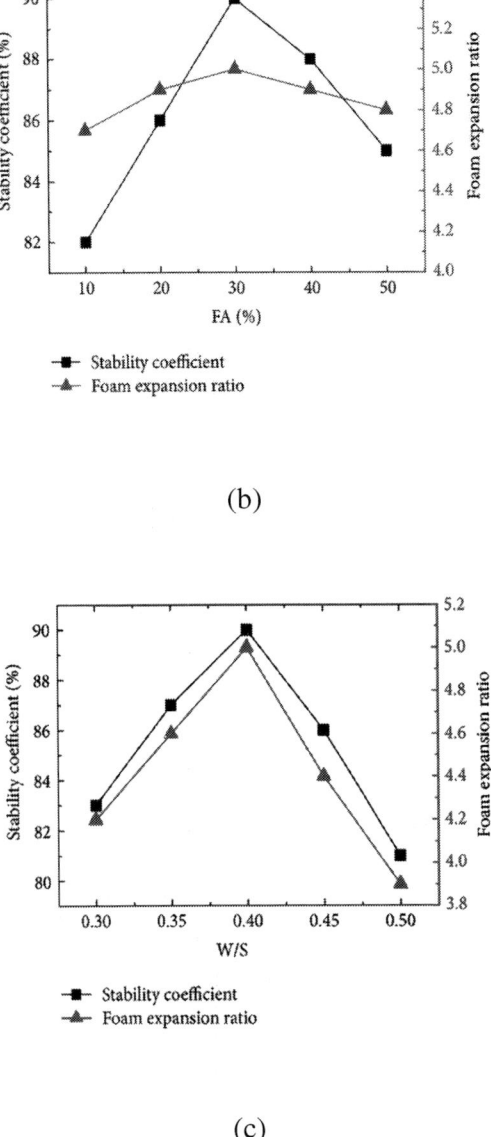

(b)

(c)

Figure 7: The change curves of stability coefficient and foam expansion ratio with different factors. (a) Independent variable is FV and constants are 30 wt.% of FA and W/S of 0.4; (b) Independent variable is FA and constants are 8 V of FV and W/S of 0.4; (c) independent variable is W/S and constants are 8 V of FV and 30 wt.% of FA.

From Figure 7(a), it can be seen that when FV increased, foam expansion ratio and stability coefficient of ISF show different variation trends. As the aqueous foam volume increases, the slurry system becomes more disperse and the film becomes thinner, which lead to the bursting of liquid film even if the drainage volume is not big. Besides, cement and fly ash particles cannot form a continuum and the setting and the hydration are slowed down. So, foam stability decreases as a function of increasing aqueous foam volume. When ISF is used in field, ISF is required with high foam expansion ratio and desired stability coefficient. But, in fact, these two targets cannot be achieved simultaneously. Therefore, we expect that, under the limit of foam expansion ratio which is not less than 4 V according to the technological requirements, the stability coefficient should be improved as high as possible.

To reduce cost, we use a small quantity of fly ash to replace cement. Figure 7(b) shows the variation in foam expansion ratio and stability coefficient with FA. The foam expansion ratio and stability coefficient both reach the maximums when the FA is 30%. This phenomenon can be explained as follows. When the cement particles are irregular geometry, there are many spherical particles (glass beads) in fly ash. Glass beads function like ball bearing reducing the friction among cement particles and increasing the liquidity of foam slurry, thus making bubbles disperse evenly. But the hydration velocity of fly ash is slower than cement. If the fly ash replacement level is too high, it can cause reduction of the early hydration products and rupture of bubbles and reduce the stability coefficient of ISF.

It is observed from Figure 7(c) that, with the increase of water cement ratio, foam expansion ratio and stability coefficient exhibit the same change trend. A possible reason for this is that, at a too low W/S level, cement hydration consumes the water of foam, leading to bubble rupture and foam slurry instability. However, when the W/S is too large, solid particles may sink and foam can float upward, which causes the uneven component of foam slurry and affects the stability of ISF.

The Enhancement of the Stability

The maximum stability coefficient is 90% based on the results of 125 groups of experiments. There are certain changes in its internal structure of foam fluid during the solidification from the fresh state. For a more in-depth study on the changes in the internal structure of the bubble, the fresh state of foam fluid (Figure 8) was observed by optical microscopic system.

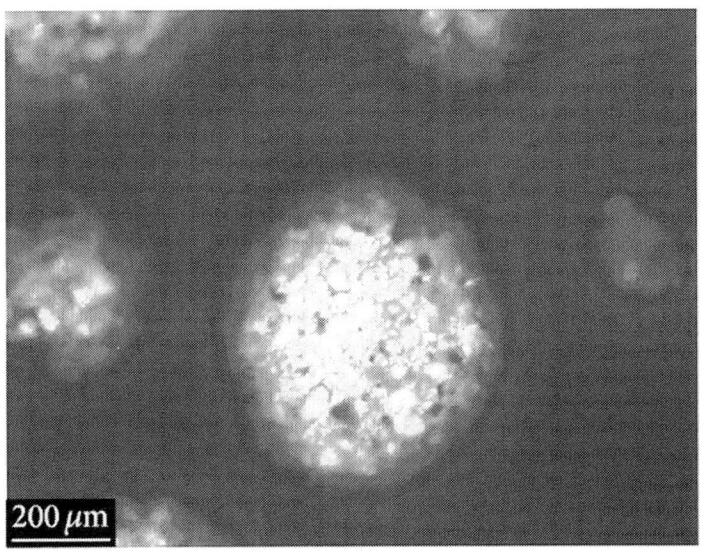

Figure 8: The fresh state of foam fluid.

Figure 8 shows that there are two distinct cases with respect to the cement particles' location. Most of the particles are present only inside the film and just a few particles are firmly attached to the film surface. In the first case, solid particles at sufficiently high concentration can form a layered structure inside the thinning film and thus stabilize it by the so called oscillatory structural force. In the second case, a few particles irreversibly adsorb at the gas-liquid interface and significantly increase the interfacial elasticity needed to prevent the film rupture and bubble coalescence. The foam stability has been quantitatively assessed by the particle hydrophobicity measured in terms of the contact angle, θ, which is related with the energy, G, required to remove the small particles (radius being R_s) from the interface by the following [26]:

$$G = \pi R_s^2 \gamma_{LG}(1 - \cos \theta)^2. \tag{7}$$

According to Binks, the optimum contact angle for foam stabilization is about 90°, as at this value the energy to remove the particle from the interface has the highest value. Experimentally, the optimum contact angle interval, ensuring the highest foam stability, was found between 40 and 70° [27] and 75 and 85° [28] (see also results in [29]). Based on (6), our measurements give a contact angle of 166° and 78° for the particles in water and surfactant solution, respectively. Therefore, the aqueous foams can be stabilized by solid particles. The adsorption of CTAB and LA molecules on the surfaces of the particles changes their hydrophobicity. The partially hydrophobic particles are able to attach to the interfaces, which play a crucial role in the high foam stability reported here [30]. For further investigation, the bursting process of an unstable bubble was shown in Figure 9.

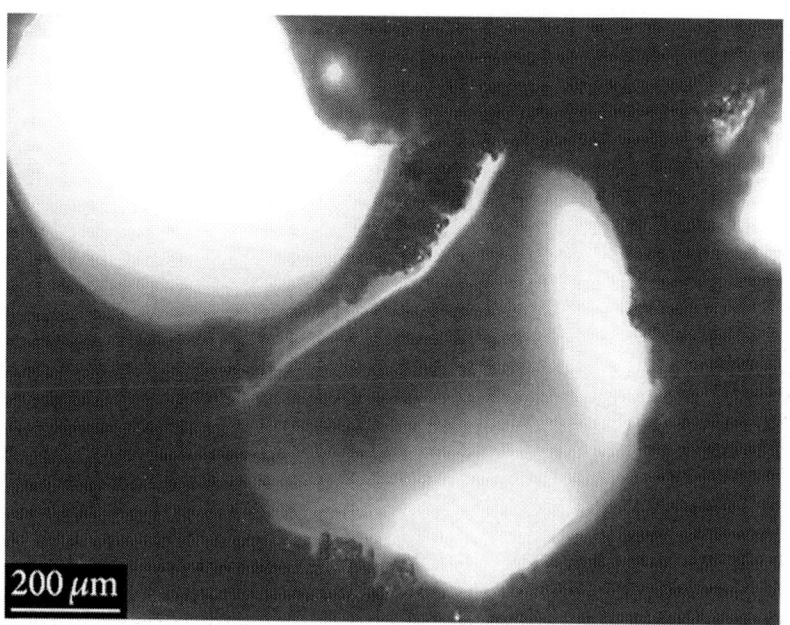

200 μm

(a)

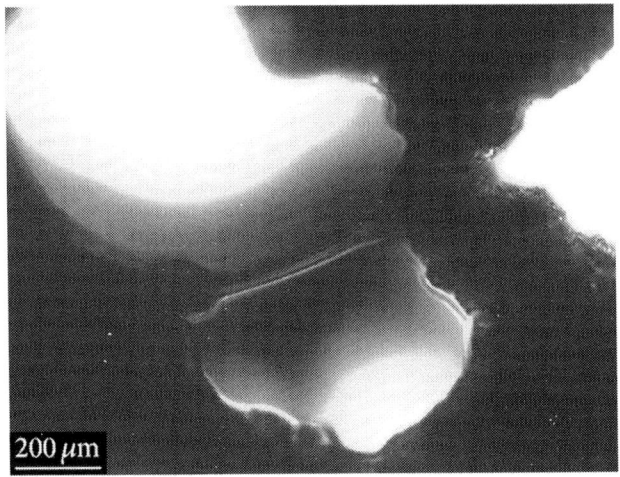

(b)

Figure 9: The bursting process of an unstable bubble.

From Figure 9(a) to Figure 9(b), this phenomenon was called limited coalescence and was observed with emulsion stabilized by the same type of particles [31]. After a drainage period the site where the liquid drained is clear (as compared with the dispersions which are turbid), and the foam evolves little with time. If initially after creation the bubble surfaces are not sufficiently covered by particles, upon coalescence, the surface to volume ratio of the created bubbles decreases and hence eventually the coalescence proceeds [32, 33].

Based on the previous analysis, the apparent high stability against disproportionation is the most significant result, even considering the coagulated nature of the particles. Also, as with foam fluid, partial coagulation of particle networks on the surfaces of the bubbles is found to be advantageous for stability. It should be noted that the rate of drainage from the bubble wall is much faster than the rate of precipitation of the hydration products. So promoting the formation of hydration products is the correct way to delay and stop the burst of bubbles.

The Dynamic Changes of Bubble Wall after Adding Calcium Fluoroaluminate

Accelerators influence the rate of cement hydration, leading to particles with a high degree of internetworking against disproportionation and to occurrence of greater retardation. So we conduct experiments on the concentration of calcium fluoroaluminate ($11CaO \cdot 7Al_2O_3 \cdot CaF_2$) on the foam stability coefficient as shown in Figure 10. Foam stability first increases and then decreases with the content increase of $11CaO \cdot 7Al_2O_3 \cdot CaF_2$ and the maximum stability is 95% under the value of concentration being 12%.

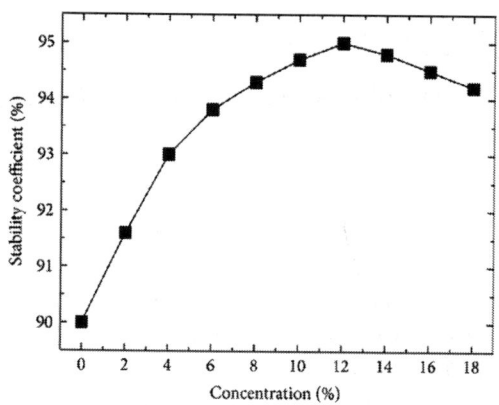

Figure 10: The foam stability coefficient versus concentration of accelerators.

In cement-based materials (e.g., ISF), the transformation process from a paste phase into a solid phase can be understood from the properties of their constituents. When $11CaO \cdot 7Al_2O_3 \cdot CaF_2$ is added to ISF system, Al_2O_3, coming from the admixture, could react with gypsumto form immediately ettringite crystals ($[Ca_2 (Al, Fe) (OH)_6]_2 \cdot X_3 \cdot nH_2O$), which will attach to the particle surface. At the same time, the consumption of gypsum accelerates the pace of tricalcium silicate ($3CaO \cdot SiO_2$) hydration, forming a small amount of fibrous C–S–H filling among the cement particles. The chemical reaction consists in the transformation of $11CaO \cdot 7Al_2O_3 \cdot CaF_2$ into $[Ca_2 (Al, Fe) (OH)_6]_2 \cdot X_3 \cdot nH_2O$ via a dissolution precipitation process by (8). The dynamic changes of bubble wall in the stabilization and solidification process were shown in Figure 11. Consider

$$3\left(11\mathrm{CaO} \cdot 7\mathrm{Al}_2\mathrm{O}_3 \cdot \mathrm{CaF}_2\right) + 33\mathrm{CaSO}_4 + 382\mathrm{H}_2\mathrm{O}$$

$$\longrightarrow 11\left(3\mathrm{CaO} \cdot \mathrm{Al}_2\mathrm{O}_3 \cdot 3\mathrm{CaSO}_4 \cdot 32\mathrm{H}_2\mathrm{O}\right)$$

$$+ 3\mathrm{CaF}_2 + 10\left(\mathrm{Al}_2\mathrm{O}_3 \cdot 3\mathrm{H}_2\mathrm{O}\right) \tag{8}$$

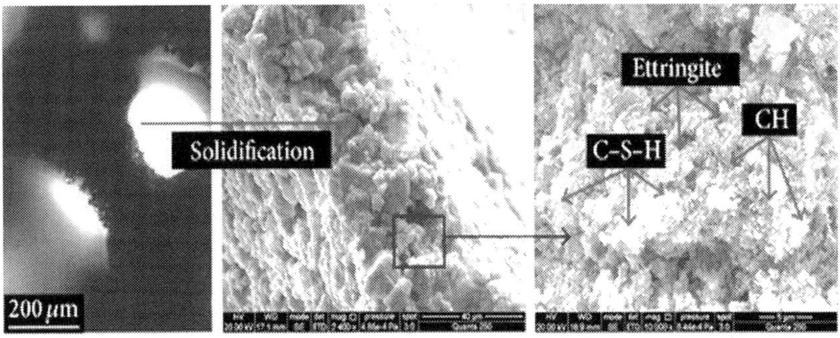

Figure 11: The SEM images of bubble wall.

In the SEM image obtained from the sample after solidification, the evolution of the primary cement hydration products is obvious. We can observe the formation of ettringite as rod-like crystals massively fill capillary pores. Surface products such as C–S–H gel can be observed as the major ISF microstructure component. CH as a pore product with a polycrystalline shape is another dominant cement hydration product. The SEM shows that the cement and fly ash particles are more connected and cement hydration products completely surround the particles.

CONCLUSIONS

- This paper presents the manufacturing process of ISF which consists of mixing the composite slurry, preparing aqueous foam, and mixing them with accelerator. The foam generator can produce homogenous and dense aqueous foams due to the turbulent eddy which is formed and enhanced by the porous medium. A large number of foam fluids are formed by the self-made mixer in which turbulence and vortex were generated, and then aqueous foams were added stepwise to slurry.

- The aqueous foam with expansion ratio of 28 and 30 min drainage rate of 13% was obtained as a function of 2.5 wt. % SDS and 2 wt. % LA. The effects of FV, FA, and W/S on stability coefficient and foam expansion ratio were studied. And the results show that the optimum values of foam expansion ratio and stability coefficient were 5 V and 90%, respectively, by value of FV being 8 V, FA being 30%, and W/S being 0.4.
- The adsorption of CTAB and LA molecules on the surfaces of the particles changes their hydrophobicity with the contact angle from 166° to 78°. The mechanism concerning accelerating the hydration and reducing the drainage was proposed and verified based on the analysis of dynamic change of bubble wall.
- At last, ISF with stability coefficient of 95% and foaming expansion ratio of 5 was fabricated, which could sufficiently satisfy field process requirements of air sealing and thermal insulation.

ACKNOWLEDGMENTS

This work was supported by the National Natural Science Foundation of China (no. U1361213), the Fundamental Research Funds for the Central Universities (CUMT, 2014YC04), and the independent study projects of State Key Laboratory of Coal Resources and Mine Safety (SKLCRSM13X04).

REFERENCES

1.   G. B. Stracher and T. P. Taylor, "Coal fires burning out of control around the world: Thermodynamic recipe for environmental catastrophe," International Journal of Coal Geology, vol. 59, no. 1-2, pp. 7–17, 2004. · ·

2.   C. Kuenzer and G. B. Stracher, "Geomorphology of coal seam fires," Geomorphology, vol. 138, no. 1, pp. 209–222, 2012. · ·

3.   M. A. Engle, L. F. Radke, E. L. Heffern et al., "Gas emissions, minerals, and tars associated with three coal fires, Powder River Basin, USA," Science of the Total Environment, vol. 420, pp. 146–159, 2012. · ·

4. F. Zhou, B. Shi, Y. Liu, X. Song, J. Cheng, and S. Hu, "Coating material of air sealing in coal mine: clay composite slurry (CCS)," Applied Clay Science, vol. 80-81, pp. 299–304, 2013. · ·

5. B. Taraba, Z. Michale, V. Michalcová, T. Blejcha , M. Bojko, and M. Kozubková, "CFD simulations of the effect of wind on the spontaneous heating of coal stockpiles," Fuel, vol. 118, no. 2, pp. 107–112, 2014. ·

6. T. R. Jolley and H. W. Russell, "Control of fires in inactivecoal deposits in Western United States, including Alaska, 1948–1958," Information Circular U.S. Bureau of Mines 7932, 1959.

7. J. J. Feiler and G. J. Colaizzi, IHI Mine Fire Control Project Utilizing Foamed Grout Technology, Rifle, Colorado Bureau of Mines, United States Department of the Interior Research Contract Report 14320395H0002, 1996.

8. F. B. Zhou, "Application of new material as air tight coating material in entries retained at gob-sides,"Coal Safety Special Issue, pp. 97–98, 2009.

9. A. Kan and H. Houde, "Effective thermal conductivity of open cell polyurethane foam based on the fractal theory," Advances in Materials Science and Engineering, vol. 2013, Article ID 125267, 7 pages, 2013. ·

10. S. K. Lim, C. S. Tan, O. Y. Lim, and Y. L. Lee, "Fresh and hardened properties of lightweight foamed concrete with palm oil fuel ash as filler," Construction and Building Materials, vol. 46, pp. 39–47, 2013. · ·

11. I. S. Ranjani and K. Ramamurthy, "Relative assessment of density and stability of foam produced with four synthetic surfactants," Materials and Structures, vol. 43, no. 10, pp. 1317–1325, 2010. · ·

12. E. Carey and C. Stubenrauch, "Free drainage of aqueous foams stabilized by mixtures of a non-ionic (C_{12}DMPO) and an ionic (C12TAB) surfactant," Colloids and Surfaces A: Physicochemical and Engineering Aspects, vol. 419, pp. 7–14, 2013. · ·

13. U. T. Gonzenbach, A. R. Studart, E. Tervoort, and L. J. Gauckler, "Stabilization of foams with inorganic colloidal particles," Langmuir, vol. 22, no. 26, pp. 10983–10988, 2006. · ·

14. T. N. Hunter, R. J. Pugh, G. V. Franks, and G. J. Jameson, "The role of particles in stabilising foams and emulsions," Advances in Colloid and Interface Science, vol. 137, no. 2, pp. 57–81, 2008. · ·

15. G. N. Sethumadhavan, A. D. Nikolov, and D. T. Wasan, "Stability of liquid films containing monodisperse colloidal particles," Journal of Colloid and Interface Science, vol. 240, no. 1, pp. 105–112, 2001. · ·

16. K. Vijayaraghavan, A. Nikolov, and D. Wasan, "Foam formation and mitigation in a three-phase gas-liquid-particulate system," Advances in Colloid and Interface Science, vol. 123-126, pp. 49–61, 2006. · ·

17. BS EN 197-1, Cement, Composition, Specifications and Conformity Criteria for Common Cements, British Standards Institution, London, UK, 1995.

18. BS EN 450, Fly Ash for Concrete: Definitions, Requirements and Quality Control, British Standards Institution, London, UK, 1995.

19. W. E. Brewer, Durability Factors Affecting CLSM. SP 150-3, American Concrete Institute, Detroit, Mich, USA, 1994.

20. A. V. Nguyen, "Flotation," in Encyclopedia of Separation Science, I. D. Wilson, Ed., pp. 1–27, Elsevier, Amsterdam, The Netherlands, 2007.

21. S. Grandner and S. H. Klapp, "Surface charge induced freezing of colloidal suspensions," Europhysics Letters, vol. 90, no. 6, Article ID 68004, 2010. ·

22. S. I. Karakashev, E. D. Manev, R. Tsekov, and A. V. Nguyen, "Effect of ionic surfactants on drainage and equilibrium thickness of emulsion films," Journal of Colloid and Interface Science, vol. 318, no. 2, pp. 358–364, 2008. · ·

23. M. Wang, H. Du, A. Guo, R. Hao, and Z. Hou, "Microstructure control in ceramic foams via mixed cationic/anionic surfactant," Materials Letters, vol. 88, pp. 97–100, 2012. · ·

24. A. D. Nikolov and D. T. Wasan, "Ordered micelle structuring in thin films formed from anionic surfactant solutions. I. Experimental," Journal of Colloid and Interface Science, vol. 133, no. 1, pp. 1–12, 1989. · ·

25. S. E. Anachkov, K. D. Danov, E. S. Basheva, P. A. Kralchevsky, and K. P. Ananthapadmanabhan, "Determination of the aggregation number and charge of ionic surfactant micelles from the stepwise thinning of foam films," Advances in Colloid and Interface Science, vol. 183-184, pp. 55–67, 2012. · ·

26. B. P. Binks, "Particles as surfactants—similarities and differences," Current Opinion in Colloid and Interface Science, vol. 7, no. 1-2, pp. 21–41, 2002. · ·

27. G. Johansson and R. J. Pugh, "The influence of particle size and hydrophobicity on the stability of mineralized froths," International Journal of Mineral Processing, vol. 34, no. 1-2, pp. 1–21, 1992. · ·

28. Y. Q. Sun and T. Gao, "The optimum wetting angle for the stabilization of liquid-metal foams by ceramic particles: experimental simulations," Metallurgical and Materials Transactions A, vol. 33, no. 10, pp. 3285–3292, 2002. · ·

29. S. W. Ip, Y. Wang, and J. M. Toguri, "Aluminum foam stabilization by solid particles," Canadian Metallurgical Quarterly, vol. 38, no. 1, pp. 81–92, 1999. · ·

30. Q. Liu, S. Zhang, D. Sun, and J. Xu, "Aqueous foams stabilized by hexylamine-modified Laponite particles," Colloids and Surfaces A Physicochemical and Engineering Aspects, vol. 338, no. 1–3, pp. 40–46, 2009. · ·

31. E. Rio, W. Drenckhan, A. Salonen, and D. Langevin, "Unusually stable liquid foams," Advances in Colloid and Interface Science, vol. 205, pp. 74–86, 2014.

32. S. Samanta and P. Ghosh, "Coalescence of bubbles and stability of foams in aqueous solutions of Tween surfactants," Chemical Engineering Research and Design, vol. 89, no. 11, pp. 2344–2355, 2011. · ·

33. W. Kracht and H. Rebolledo, "Study of the local critical coalescence concentration (l-CCC) of alcohols and salts at bubble formation in two-phase systems," Minerals Engineering, vol. 50-51, pp. 77–82, 2013. · ·

Citations

CHAPTER 1

Sachin C. Gondhalekar and Sanjeev R. Shukla, "Recovery of Ga(III) by Raw and Alkali Treated Citrus limettaPeels," International Scholarly Research Notices, vol. 2014, Article ID 968402, 10 pages, 2014. doi:10.1155/2014/968402.

CHAPTER 2

Lean Poh Goh, Khairunisak Abdul Razak, Nur Syafinaz Ridhuan, Kuan Yew Cheong,Poh Choon Ooi, and Kean Chin Aw, Direct Formation of Gold Nanoparticles on Substrates Using a Novel ZnO Sacrificial Templated-growth Hydrothermal Approach and their Properties in Organic Memory Device, doi:10.1186/1556-276X-7-563.

CHAPTER 3

Salmah B. Karman, S. Zaleha M. Diah, and Ille C. Gebeshuber, "Raw Materials Synthesis from Heavy Metal Industry Effluents with Bioremediation and Phytomining: A Biomimetic Resource Management Approach,"Advances in Materials Science and Engineering, vol. 2015, Article ID 185071, 21 pages, 2015. doi:10.1155/2015/185071.

CHAPTER 4

Yogesh B. Patil (2013). Development of a Bioremediation Technology for the Removal of Thiocyanate from Aqueous Industrial Wastes Using Metabolically Active Microorganisms, Applied Bioremediation - Active and Passive Approaches, Dr. Yogesh Patil (Ed.), ISBN: 978-953-51-1200-6, InTech, DOI: 10.5772/56975.

CHAPTER 5

Mohd Aslam, Sumbul Rais, Masood Alam, and Arulazhagan Pugazhendi, "Adsorption of Hg(II) from Aqueous Solution Using Adulsa (Justicia adhatoda) Leaves Powder: Kinetic and Equilibrium Studies," Journal of Chemistry, vol. 2013, Article ID 174807, 11 pages, 2013. doi:10.1155/2013/174807.

CHAPTER 6

Anna Lenart-Boroń and Piotr Boroń (2014). The Effect of Industrial Heavy Metal Pollution on Microbial Abundance and Diversity in Soils — A Review, Environmental Risk Assessment of Soil Contamination, Dr. Maria C. Hernandez Soriano (Ed.), ISBN: 978-953-51-1235-8, InTech, DOI: 10.5772/57406.

CHAPTER 7

Maria K. Cherepivska and Roman V. Prihod'ko, "Sol-Gel Synthesized Semiconductor Oxides in Photocatalytic Degradation of Phenol," ISRN Physical Chemistry, vol. 2014, Article ID 724095, 7 pages, 2014. doi:10.1155/2014/724095.

CHAPTER 8

Nilisha Itankar, Viraja Bhat, Jayati Chourey, Ketaki Barve, Shilpa Kulkarni, Prakash Rao and Yogesh Patil (2013). Resource Recovery from Industrial Effluents Containing Precious Metal Species Using Low-Cost Biomaterials — An Approach of Passive Bioremediation and Its Newer Applications, Applied Bioremediation - Active and Passive Approaches, Dr. Yogesh Patil (Ed.), ISBN: 978-953-51-1200-6, InTech, DOI: 10.5772/56965.

CHAPTER 9

Botao Qin, Yi Lu, Fanglei Li, Yuwei Jia, Chao Zhu, and Quanlin Shi, "Preparation and Stability of Inorganic Solidified Foam for Preventing Coal Fires," Advances in Materials Science and Engineering, vol. 2014, Article ID 347386, 10 pages, 2014. doi:10.1155/2014/347386.

Index